AF900054

Conversion of Former BTW Facilities

NATO Science Series

A Series presenting the results of activities sponsored by the NATO Science Committee. The Series is published by IOS Press and Kluwer Academic Publishers, in conjunction with the NATO Scientific Affairs Division.

General Sub-Series

A. Life Sciences	IOS Press
B. Physics	Kluwer Academic Publishers
C. Mathematical and Physical Sciences	Kluwer Academic Publishers
D. Behavioural and Social Sciences	Kluwer Academic Publishers
E. Applied Sciences	Kluwer Academic Publishers
F. Computer and Systems Sciences	IOS Press

Partnership Sub-Series

1. Disarmament Technologies	Kluwer Academic Publishers
2. Environmental Security	Kluwer Academic Publishers
3. High Technology	Kluwer Academic Publishers
4. Science and Technology Policy	IOS Press
5. Computer Networking	IOS Press

The Partnership Sub-Series incorporates activities undertaken in collaboration with NATO's Partners in the Euro-Atlantic Partnership Council – countries of the CIS and Central and Eastern Europe – in Priority Areas of concern to those countries.

NATO-PCO-DATA BASE

The NATO Science Series continues the series of books published formerly in the NATO ASI Series. An electronic index to the NATO ASI Series provides full bibliographical references (with keywords and/or abstracts) to more than 50000 contributions from international scientists published in all sections of the NATO ASI Series.
Access to the NATO-PCO-DATA BASE is possible via CD-ROM "NATO-PCO-DATA BASE" with user-friendly retrieval software in English, French and German (© WTV GmbH and DATAWARE Technologies Inc. 1989).

The CD-ROM of the NATO ASI Series can be ordered from: PCO, Overijse, Belgium.

Series 1: Disarmament Technologies – Vol. 21

Conversion of Former BTW Facilities

edited by

Erhard Geissler
Max-Delbrück-Center for Molecular Medicine,
Berlin-Buch, Germany

Lajos Gazsó
National Research Institute for Radiobiology and Radiohygiene,
Budapest, Hungary

and

Ernst Buder
Max-Delbrück-Center for Molecular Medicine,
Berlin-Buch, Germany

Kluwer Academic Publishers

Dordrecht / Boston / London

Published in cooperation with NATO Scientific Affairs Division

Proceedings of the NATO Advanced Research Workshop on
Conversion of Former BTW Facilities: Development & Production of Prophylactic, Diagnostic & Therapeutic Measures for Countering Diseases
Budapest, Hungary
5–9 November, 1997

A C.I.P. Catalogue record for this book is available from the Library of Congress

ISBN 0-7923-5249-1

Published by Kluwer Academic Publishers,
P.O. Box 17, 3300 AA Dordrecht, The Netherlands.

Sold and distributed in North, Central and South America
by Kluwer Academic Publishers,
101 Philip Drive, Norwell, MA 02061, U.S.A.

In all other countries, sold and distributed
by Kluwer Academic Publishers,
P.O. Box 322, 3300 AH Dordrecht, The Netherlands.

Printed on acid-free paper

All Rights Reserved
© 1998 Kluwer Academic Publishers and copyrightholders
as specified on appropriate pages within.
No part of the material protected by this copyright notice may be reproduced or utilized in any form or by any means, electronic or mechanical, including photocopying, recording or by any information storage and retrieval system, without written permission from the copyright owner.

Printed in the Netherlands

Table of Contents

Annexes

Introduction

The development, production, stockpiling and use in war of biological and toxin weapons are prohibited by international law. Although not explicitly stated, the two treaties outlawing such activities, the Geneva Protocol of 1925 and the Biological and Toxin Weapons Convention of 1972, prohibit the continuation of activities previously performed in Biological and Toxin Weapons facilities not justified for prophylactic, protective or other peaceful purposes.

Because conversion and other means of cessation of former BTW facilities are not explicitly addressed in the treaties mentioned above the problems involved in conversion of BTW facilities have thus far only been discussed marginally in the open literature. In times of increased awareness of the danger of biological and toxin warfare (including the increased danger of terrorist use of biological and toxin weapons) it seemed necessary to us to invite experts from different parts of the world to discuss the pros and cons of conversion and the problems involved. It also became obvious to us that the conversion of former BTW facilities should be discussed with respect to the necessity of peaceful international cooperation in areas related to the Biological and Toxin Weapons Convention.

An additional reason to discuss matters of peaceful cooperation is that cooperation is explictly requested by Article X of the Biological and Toxin Weapons Convention. Because numerous States Parties to this Convention complain about insufficient implementation of that very article, the mandate of the Ad Hoc Group of governmental experts presently engaged in elaborating a protocol to the Biological and Toxin Weapons Convention includes the consideration of "specific measures designed to ensure effective and full implementation of Article X". It might be advantageous, therefore, to discuss conversion of former BTW facilities also in terms of implementation of Article X.

When we expressed our views about the necessity to enhance international activities towards conversion at a NATO Advanced Research Workshop on "Developing Future Scientific Initiatives in Disarmament Technologies", which took place in Berlin, Germany, in October 1996, we were encouraged by the participants and by NATO's Scientific and Environmental Affairs Division to organize a NATO Advanced Research Workshop for that purpose. With generous support from NATO the conference was held at the "Frédéric Joliot-Curie" National Research Institute for Radiobiology and Radiohygiene in Budapest, Hungary, on 5 to 9 November, 1997. The workshop was attended by 40 participants from three NATO countries (Germany, UK and US) and four Cooperation Partner countries (Hungary, Poland, Russia and Ukraine) was well as Sweden.

Some major contributions presented at the workshop as well as an additional paper contributed by Anthony Rimmington, who was unable to participate in the meeting, are presented in this book.

The editors would like to take this opportunity to thank NATO's Scientific and Environmental Affairs Division for the support given both to organizing the conference and to the publication of this book. We are especially grateful to Ms. Nancy Schulte from that Division for her continuous support.

The editors express their gratitude also to Prof. G. J. Köteles, Dr. Gy. A. Jánoki and Ms. Elisabeth Tóth from the "Frédéric Joliot-Curie" Institute for Radiobiology and Radiohygiene in Budapest and to Ms. Iris Hunger of the Bioethical Research Group of the Max Delbrück Center for Molecular Medicine for their assistance. We thank also Dr. Harry Smith, editorial office of the Max Delbrück Centre, for editorial help.

Berlin and Budapest, April 1998

Ernst Buder Lajos Gazso Erhard Geissler

CONVERSION OF FORMER BTW FACILITIES - QUESTIONS AND PROBLEMS

ERNST BUDER and ERHARD GEISSLER
Max-Delbrück-Center for Molecular Medicine,
Berlin-Buch, Germany

1. Introduction

As already noted, this book is a result of the NATO Advanced Research Workshop "Conversion of Former Biological Weapons Facilities: Development and Production of Prophylactic, Diagnostic and Therapeutic Measures for Countering Diseases" held in Budapest 5-9 November 1997. Since only a few authors have dealt in the past with the problems of conversion of BTW facilities [1-3] a set of exploratory questions was compiled and distributed among participants of the NATO workshop in advance. This paper is a short summary of the response to the questionnaire.

1.1. THE NECESSITY TO STOP THE ACTIVITIES OF BTW FACILITIES

The danger of a biological warfare (BW) has existed at least since World War I (WWI). In modern times it was Germany who introduced the hostile use of bacteria for sabotage purposes against animals in 1915-17 [4]. More or less in response to the German WWI-activities France, the Soviet Union, Japan, Italy, Hungary, the United Kingdom, Canada, and the USA implemented variably extensive BW and toxin warfare (TW) (defense) programmes before and during World War II (WWII) [5]. The most extensive BTW programme was performed by Japan, who went so far as to use biological weapons in WWII. The Anglo-Americans together with their Canadian allies were prepared to use BW for retaliation should Germany use such weapons first. In fact, however, biological warfare did not take place in the European theatre primarily due to Germany's reluctance to use such weapons [6].

After WWII the danger of biological warfare persisted and even increased during the Cold War as the two major superpowers and some of their allies carried out BTW programmes. The US programme was stopped unilaterally in 1969 and 1970 by decisions of President Nixon [7,8] that became instrumental in triggering the adoption of the Biological and Toxin Weapons Convention (BTWC) in 1972. Consequently the BTW facilities of the USA and the UK were converted (see Pearson, and Robertson, this volume).

E. Geissler et al. (eds.), Conversion of Former BTW Facilities, 1-10.
© 1998 *Kluwer Academic Publishers. Printed in the Netherlands.*

In spite of the existence of the Convention, BTW activities were not terminated globally. The Soviet Union continued to run an offensive BTW programme [9] until it was officially stopped by President Yeltzin in 1992 [10, 11]. Regarding the eventual conversion of the former Soviet BTW facilities contradictory views are expressed (see, for example, Leitenberg and Sandakhchiev, this volume). According to a recent evaluation of the US Department of Defense Russia might "be continuing some research related to biological warfare" [12]. A Soviet defector claimed that Russia continues BTW activities [13]. It is even alleged that Russia has recently developed a new biological weapon in the form of anthrax bacteria genetically modified for resistance to all known antibiotics [14] and that the former BTW facility located in Ekaterinburg has been reverted again [15].

After the Gulf War it became evident that Iraq had also executed an extensive BTW programme [16]. At least some of Iraq's BTW facilities were destroyed during and after the war, but such great concern has persisted that Iraq is continuing a BTW programme that in February 1998 the US Administration considered military action to terminate the activities of suspected facilities and to destroy any residual hidden stockpiles of biological and chemical weapons in Iraq.

In addition to Iraq some 20 countries are suspected of having or developing weapons of mass destruction including biological weapons [12, 17, 18].

There is, moreover, a growing risk that BTW agents could also be used by terrorist groups [19]. Investigations triggered by the CW attack by the Aum Shinriky Sect in the subway system of Tokyo in March 1995 indicated that the sect had worked on biological weapons also [20].

1.2. THE NECESSITY TO AGREE ON A COMMON LANGUAGE

Discussions on conversion of former BTW facilities suffer, *inter alia*, from the lack of relevant definitions. For example: what is a "BTW facility"? What is "conversion"? Do all people mean the same thing when they use the term "conversion"? And: is conversion a sufficient means of terminating institutional BTW activities? (see Geissler, this volume). Finally: what are the similarities and differences in the conversion of BTW facilities in comparison to conversion of nuclear or chemical weapons production facilities?

To contribute to a common understanding of the terms used in our discussion we have drafted a list of relevant terms and definitions (see the annex to this book). The terms used throughout this chapter are defined in that annex. If not otherwise stated throughout this chapter the term "conversion" is used for conversion of BTW facilities, i.e., facilities involved in offensive military BTW activities. BTW defence facilities, i.e. facilities where activities *not* prohibited by the Biological and Toxin Weapons Convention (BTWC) are performed, are considered in this chapter only if they are the product of partial conversion.

The major difference between BTW facilities and facilities involved in the development and production of conventional, chemical or nuclear weapons arises from the very nature of BTW agents. BW agents, be they viruses, bacteria, fungi, anthropods as well as TW agents, are "dual-threat agents" (DTAs) [21]. First, they are naturally pathogens, and, second, they can be used for hostile purposes to cause disease or death in humans, animals or plants. Research on the structure and function of DTAs as well as development of measures for their detection and differential diagnosis and research and development (R&D) to combat their pathogenic properties is necessary, therefore, to both civilian public health and military purposes. Since R&D on DTAs can be carried out for civilian and military purposes, and since military R&D on DTAs can be performed both for defensive and offensive purposes, we are confronted with dual-use R&D which depends on dual-use technology and equipment. It is very often extremely difficult, therefore, to differentiate between permitted and prohibited activities with regard to DTAs. Tucker has called this the "dilemma of the dual-use technology" [22]. In fact, it is "dual-threat-and-dual-use dilemma". This dilemma makes it difficult to decide, first, whether it is sufficient to stop the activity of a given BTW facility by conversion, and, second, to verify whether conversion has been complete.

2. Possibilities for cessation of prohibited activities

Although not explicitly mentioned in the Biological and Toxin Weapons Convention it follows implicitly from the first two articles of the BTWC that all former BTW facilities should be ceased. Article I requests[1] that

> "Each State Party to the Convention undertakes never in any circumstances to develop, produce, stockpile or otherwise acquire or retain:
> 1. Microbial or other biological agents, or toxins whatever their origin or method of production, of types and in quantities that have no justification for prophylactic, protective or other peaceful purposes;
> 2. Weapons, equipment or means of dilvery designed to use such agents or toxins for hostile purposes or in armed conflict".

And Article II demands that:

> "Each State Party to this Convention undertakes to destroy, or to divert to peaceful purposes [...] all agents, toxins, weapons, equipment and means of delivery specified in article I of the Convention, which are in its possession or under its jurisdiction or control. [...]"

There are a variety of means for reaching these goals and terminating the functioning of BTW facilities: conversion, closing (coupled with sealing), dismantling, and destruction (see annex).

These measures have different major advantages and disadvantages. The advantages of conversion, which are described in more detail below, are that one can make use of the facility including its personnel, equipment and infrastructure for permitted purposes. At least part of the costs necessary for cessation can then be recovered directly by selling the products of the converted facilities. In addition, conversion allows for well-trained personnel to be preserved, thus limiting the danger

---

[1] The text of the BTWC is reprinted, *inter alia*, in [23].

of a "brain drain". The major disadvantage of conversion is--in terms of arms control and disarmament--that a converted facility can be totally reverted to a functioning BTW facility with ease (as has been claimed recently with respect to the Center for Military and Technical Problems of Anti-Bacteriological Defense in Ekaterinburg by the facility's retired director of personnel [15]).

The advantage of closing, complete dismantling and destruction is the preclusion of reversion. The disadvantages are that the former facility cannot be used for permitted purposes (although their might be an urgent public health demand for the availability of such facilities), that the funds necessary can not be recovered, and that there is the danger of a brain drain (see below).

A government deciding to stop BTW activities in order to comply with the BTWC and/or for other reasons must determine the appropriate measures of termination on a case-by-case basis, taking into account the market situation, capabilities of the given facility and possible financing arrangements. Economic considerations are important but should not be decisive.

As soon as an organization for the prohibition of biological and toxin weapons is established as envisioned by the Ad Hoc Group (AHG) of States Parties to the BTWC [24] a given government might consult the organization and/or its secretariat for advice as to how to terminate the activities of a given facility. After termination of their activities former BTW facilities should be subject to the compliance measures proposed by the AHG and covered by an envisaged additional protocol to the BTWC.[2] In that context one might consider whether the organization upon its formation could advise relevant states as to how to terminate the activities of a facility.

2.1. ADVANTAGES OF CONVERSION

According to Leitenberg [1] the prerequisites of conversion are: "a) a total end to offensive programmes; b) a complete turnover of the laboratories to civilian control, and no covert programmes withheld from government knowledge; and c) an end to secrecy, and complete access to all the facilities by the international scientific community".

At first sight, conversion might seem the most feasible and efficient possibility for terminating offensive BTW activities at a facility. Here, the dual-threat-and-dual-use dilemma might appear as an advantage because, at least in principle, most of the personnel, equipment and technology previously used for destructive purposes could be exploited constructivly. A BTW facility formerly involved in R&D of vaccines within the frame of an offensive BTW programme, for example, might be easily converted into a vaccine production facility preparing vaccines for defence purposes or for civilian health protection (see Geissler, this volume, for some examples). At the same time it would be especially easy to reemploy experts formerly involved in prohibited activities at the converted facility.

---

[2] Supervision of former chemical weapons production facilities according to the provision of the Chemical Weapons Convention is described by Stock (this volume).

Two different possibilities for conversion can be envisaged: Conversion into a facility totally or partially engaged in BTW defence activities ("partial conversion") and conversion into a facility engaged exclusively in civilian activities ("full conversion"). Only the first of the three prerequisites of conversion defined by Leitenberg would apply to either form conversion. Neither complete turnover of the facility to civilian control nor complete access of the facility to the international scientific community would apply to facilities partially converted to BTW defence facilities.

An additional consideration with respect to conversion is that the production of vaccines and other pharmaceuticals requires very high operational standards. This could often entail extensive retraining of personnel including qualified biochemists and microbiologists, to enable them to apply current Good Manufacturing Practice (cGMP). The layout and operation of facilities must also be regarded as critical to meeting cGMP standards. Moreover, it is vital that biologicals produced within a converted facility not be contaminated by virulent or toxic agents formerly studied in that building(s) (see Melling and Robertson, this volume). It may thus be advisable to use former BTW facilities for diagnostic activities, diagnostic production, etc., whereas production of vaccines and other biologicals for internal human use, usually requiring very high safety standards, should rely on the construction of new buildings furnished with new equipment.

2.2. CRITERIA OF FULL CONVERSION

Participants responding to our questionnaire indicated the following as criteria for full conversion:

a) The availability of an open civilian agenda with open activities, purchases, information on personnel (number, training).

b) Open discussions pertaining to activities inside respective facilities.

c) Easy access for foreign experts and scientists to attend conferences in the converted facility, to take part in scientific exchanges etc.

d) Unlimited opportunities for personnel to participate in national and international conferences.

e) Unlimited possibilities of personnell to engage in joint R&D activities with other groups within and outside of the given country.

f) Unrestricted rights to publish the results of R&D in national and international journals and to present data at national and international conferences (taking into account commercial proprietary rights).

g) Transparent financing patterns.

h) Stop of funding by the ministry of defense and/or related agencies.

i) Replacement of personnel formerly acting in top positions by civilian trained and experienced people.

Also to be expected is a possible reduction in the number of employees of the facility in accord with the new direction of R&D performed. If the facility is converted into an institution involved in R&D on diagnostics and other biologicals, the extent of reduction will depend on the epidemiological situation and the necessities of controlling infectious diseases.

It is also considered axiomatic that successful conversion programmmes must entail both technical feasiblity and economical viability in the long term (see Dando, Melling, and Robertson, this volume). Full conversion especially must include clear economic options. One economical prerequisite should be the manufacture of market-driven products. Arrangements should also be made for governmental subsidies of products during the initial stages of full conversion (whereas it is understood that partial conversion is funded by governmental agencies at least in part).

One of the most important criteria is openness of a fully converted facility for on-site inspections including access to all parts of the facility, since subjection to verfication (see below) is one of the most powerful measures for ensuring efficient conversion.

2.3. THE DANGER OF REVERSION

Independent of safety concerns, numerous experts are concerned about the danger of possible reversion of converted facilities. For this very reason former BTW facilities should be, according to Leitenberg and others, preferably converted to facilities engaged in activities unrelated to BTW issues, or terminated by closure, dismantling or destruction.

Moreover, conversion and other means of ceasing BTW activities must consider the personnel involved so as to prevent both reversion and any tendency of brain drain ("human proliferation") to other countries or groups interested in BTW programmes. There are reports, for example, that weapons experts including biologists of the former Soviet Union have accepted contracts from Iran and Libya [25]. According to a Soviet defector numerous scientists who were working in the Soviet biowarfare programme left the country and "no one knows where they are" [13].

Fortunately there are also some impressive examples of successful conversion processes, such as in Germany after WWII and in the USA and UK after president Nixon's decision and adoption of the BTWC (see Geissler, Melling, Pearson, and Robertson, this volume). In addition there are encouraging reports regarding the conversion of a former major Soviet BTW production facility in Stepnogorsk, Kazachstan (see Rimmington, this volume). These examples indicate that it is indeed possible to stop offensive BTW programmes and to work out successful programmes designed to convert or otherwise terminate the relevant facilities.

In the case of Germany, successful termination resulted from both the German defeat and the political decisions of the post-war German governments and their allies[3]. In the USA and the UK conversion was the consequence of internal political decisions. Likewise the conversion of the Stepnogorsk facility was the consequence of a corresponding decision of the President of Kazachstan.

---

[3] The Federal Republic of Germany renounced the production of chemical and bacteriological weapons in a treaty with the other members of the Westeuropean Union [26]. In the German Democratic Republic obviously no CBW activities took place, presumably as consequence of decisions taken by the Soviet government.

Similarly, in 1992, President Boris Yeltsin prohibited the conduct of any activity in Russia that would violate the BTWC [10, 11]. But, as already mentioned above, doubts have been expressed (not only in Western countries) with respect to the comprehensive and complete implementation of the Russian President's order. With the possible exception of the former Institute of Molecular Biology in Koltzovo, Novosibirsk region (see Sandakhchiev and Howson, this volume), what concrete steps have so far been taken in Russia and other states of the Community of Independent States in order to terminate BTW activities remains largely unknown. It is thus obvious that cessation of former BTW facilities will depend not only on the supreme leader of a given country, but also on the support of governments, administrations, parliaments, security services and, last but not least, by the armed forces. These institutions, at the same time, should feel responsible for overseeing the essation processes and for verifying their final outcome.

2.4. PREVENTION OF REVERSION

Independent of which action is taken to bring the activities of a BTW facility to a halt measures are necessary to prevent reversion. Initiatives to elaborate a general verification system--a major target of which should be converted facilities--are underway in the framework of strengthening the BTWC (see Pearson's concluding chapter, this volume). It is essential to make cessation processes as transparent as possible to demonstrate compliance, to eliminate (possibly unfounded) suspicions, especially if a BTW facility is converted, and to provide confidence. Measures of varying intrusiveness should be established if a BTW facility is converted to a BTW defence facility or to a facility solely involved in civilian activities.

Openness and transparency are crucial elements in all processes of cessation. Measures adopted to verify cessation, moreover, should include complete access to the closed or converted facility. Investigation regimes elaborated for inclusion into an additional protocol to the BTWC and destined to evaluate the activities of declared military facilities should apply to former BTW facilities, regardless of whether they are used as BTW defence facilities or converted to facilities completely engaged in civilian activities. Because of the danger of reversion, on-site inspections without the right of refusal are an absolute necessity.

Correspondingly, former BTW facilities including those fully converted, should be requested to submit reports on activites, agents studied, equipment used etc. The latter should provide an open production schedule in order to justify their activities in real economic terms.

Additional supporting conditions and prerequisites for limiting the danger of reversion include, *inter alia*, exact descriptions of the proposed means and final results of conversion; appointment of civilian trained experts for leading positions in converted facilities; exchange programmes with scientists from foreign countries; and education programmes for people working in relevant fields as well as for the general public aimed at raising awareness of both problems and international efforts to ban biological and toxin weapons.

3. Financing Cessation

Experiences in successful conversion especially in the USA and UK (see Pearson and Robertson, this volume) demonstrate that conversion and other processes aimed at the cessation of BTW facilities are time-consuming and can be quite expensive[4].

The costs of cessation will vary and will depend on the form of cessation applied. If a former BTW facilty is to be fully converted to be used, for example, for medical diagnostic purposes, for development and production of diagnostics or for animal food production, the cost will not be so high. If it is decided, however, to convert into a facility producing vaccines, antisera, monoclonal antibodies and similar biologicals for internal medical use, then it is advisable to construct a completely new facility and to destroy the former BTW facility for the reasons discussed above. These issues are explicitly described by Melling and Robertson (this volume).

Financing civilian research and development at fully converted sites, for instance for medicine, biotechnology and pharmacy, should be market driven. There should additionally be a three-to five-year window of opportunity for fully converted facilities to become self-sustained and profitable. To attract funds for conversion there must be clear agenda as to the kinds of products that will be produced and the markets areas to be targeted.

Because financing could present a limitation on conversion, it will be first necessary that external financial support from national and international sources be available for cessation measures. Main sources currently providing funds for conversion and related measures include: the International Science and Technology Center, Moscow (see Compton, this volume); the INTAS Programme (founded by the European Community in 1993 as "Non Profit International Association" for supporting of R&D projects of scientists from Russia and the other states of the Community of Independent States [CIS]); the TACIS (Technical Assistance to the Community of Independent States) Programme, founded by the European Commission, supports projects from former Soviet military-industrial complex; and the Nunn Lugar Fund, which supports project from former military R&D of the CIS states, especially from the nuclear field. Additional sources proposed by the participants of the Budapest workshop include, *inter alia*: the Soros Fund, an International Science Foundation founded by Georges Soros; and the National Research Council of the United States which offers grants to American specialists to establish new research partnerships with colleagues from Central/Eastern Europe and the CIS States. International organizations such as WHO, UNIDO and UNESCO might also support conversion processes at least in part.

The view has been expressed that privatization of BTW facilities will be easiest in Western countries. Nevertheless, conflicts of interest may arise with regards to state and international desires to convert former BTW facilities on the one hand and the market interest of the biochemical/pharmaceutical industry on the other.

---

[4] Unfortunately the intention of the editors to include a chapter in this volume dealing with the funds necessary for conversion of former Soviet BTW facilities could not be realized.

In addition, new ways have to be sought to fund conversion and other cessation measures because of a lack of financial sources from national or international institutions and organisations like the UN and the EU. One possibility might be the enhancement of cooperative interests of biotechnological and pharmaceutical industries. Joint international work in the field of development and production of prophylactic, diagnostic and therapeutic measures for countering diseases would be very helpful for transparency and the successful conversion of former BTW facilities.

4. Final remarks and conclusions

Cessation of the activities of former BTW facilities by conversion and other means is necessary and is regarded as one of the most important confidence-building and security measures in the field of BTW arms control. Admittedly, cessation of former BTW facilities will be difficult since it depends not only on the the political readiness of governments and relevant administrations where such facilities are located, but is also time-consuming and requires enormous amounts of money. Nevertheless, cessation of such facilities by conversion and other means is in progress as several contributors to this book demonstrate.

One of the most important problems of cessation, especially if performed as conversion, is the prevention of reversion. The activities of the Ad Hoc Group of States Parties (AHG) presently drafting an additional protocol might contribute largely to the development of measures sufficient to reduce that danger. Measures developed by the AHG in order to ensure compliance with the BTWC might also be adaptable to the task of verifying conversion and limiting the danger of reversion (see Pearson, final chapter of that volume).

With respect to the support that the biotechnological and pharmaceutical industry could provide to conversion activities we wonder whether the people and bodies directing these enterprises are--with the exception of those companies already involved in joint projects with facilities of the former Soviet Union mentioned in several chapters of this book--aware of the advantages of a possible cooperation with former BTW facilities in the process of their full conversion. Presumably the industry needs more information on the possibilities offered by such facilities. Perhaps holding an international meeting between BTW experts interested in conversion and decision-makers from biotechnological, pharmaceutical and related companies would be very helpful.

5. Acknowledgements

Our work was generously supported by the Deutsche Forschungsgemeinschaft (E.B.) and the Volkswagen-Stiftung (E.G.). We thank all participants of the Budapest workshop who responded to the questionnaire, and we are grateful to Iris Hunger for helpful comments.

6. References

1. Leitenberg, M. (1994) The conversion of biological warfare research and development facilities to peaceful use, in E. Geissler and J.P. Woodall (eds.) *Control of Dual - Threat Agents: The Vaccines for Peace Programme*, SIPRI Chemical & Biological Warfare Studies **15**, Oxford University Press, Oxford, pp. 77-105.
2. Harrington, A.M. (1995) Redirecting Biological Weapons Enterprises: Realities and Opportunities in the Former Soviet Union, *Chemical Weapons Convention Bulletin* **29**, pp. 2-5.
3. Rimmington, A. (1996) From Military to Industrial Complex? The Conversion of Military Microbiological Facilities in the Russian Federation, *Contemporary Security Policy* **17**, pp.81-112.
4. Geissler, E. (1997) Anwendung von Seuchenmitteln gegen Menschen nicht erwünscht", *Militärgeschichtliche Mitteilungen*, **56**, pp. 107-55.
5. Geissler, E. and van Courtland Moon, J.E. (1998) *The Genesis of Germ Warfare: The Development and Use of Biological and Toxin Weapons from the Middle Ages to the End of World War II.* SIPRI Chemical & Biological Warfare Studies **17**, Oxford University Press, in press.
6. Geißler, E. (1998) Biologische Waffen - nicht in Hitlers Arsenalen. Biologische und Toxin-Kampfmittel in Deutschland von 1915 bis 1945. Lit-Verlag, Münster.
7. [Nixon, R.] (1970) Statement on chemical and biological defense policies and programs, November 25, 1969, in *Public Papers of the President of the United States: Richard Nixon, 1969*, U.S. Government Printing Office. Washington, DC, p. 968
8. USA (1970) U.S. renounces use of toxins as a method of warfare. *US Department of State Bulletin* **62**, no. 1601, pp. 266-67.
9. Russian Federation (1992), CBM Report, 3 July, United Nations DDA/4-92/BWIII/ Add. 3, 22 September, pp. 32-90.
10. Anonymus (1992) Yeltsin signs decree on biological weapons. *FBIS-SOV*-92-074, 16 April, p. 3.
11. AP (1992) Yeltsin commits to germ warfare ban. *Washington Post* 17 April, p. A28.
12. Office of the Secretary of Defense (1997), *Proliferation: Threat and Response.* US Department of Defense, Washington, DC., November.
13. Preston, R. (1998) Annals of warfare. The bioweaponeers. *The New Yorker*, 9 March, pp. 52-65.
14. *Jane's Land Based Air Defense 1997-98*, foreword, quoted in *Chemical and Biological Weapons Convention Bulletin* no. 36, p. 23.
15. Quinn-Judge, P. (1998) The breeding of death. Fears grow that a secret Russian germ warfare laboratory is once again making biological weapons. *Time* **151**, no. 7, pp. 22-24.
16. Zilinskas, R.A. (1997) Iraq's biological weapons: The past as future? *Journal of the American Medical Association* **278**, pp. 418-424.
17. Reed, J. (1993) *Defence Exports: Current Concerns.* A Jane's Special Brief. Jane's Consultancy Services, Coulsdon, Surrey.
18. Dando, M.R. and Pearson, G.S.(1997) Conrolling Biological Weapons. The Fourth Review Conference of the Biological and Toxin Weapons Convention: issues, outcomes, and Unifinished Business, *Politics and the Life Sciences* **16**, pp. 105-126.
19. Roundtable: Chemical/Biological Terrorism: Tucker, J.B. (1996) Chemical/biological terrorism: Coping with a new threat, *Politics and the Life Sciences* **15**, pp. 167-183; and "Commentaries", *ibid.* pp. 185-247.
20. Committee on Governmental Affairs (1995) *Global Proliferation of Weapons of Mass Destruction*, U.S. Government Printing Office, Washington, D.C.
21. Geissler, E. (1992) Vaccines for Peace: An International Program of Development and Use of Vaccines Against Dual Threat Agents, *Politics and the Life Sciences* **11**, pp. 231-243.
22. Tucker, J.B. (1994) Dilemmas of Dual-use Technology. Toxins in Medicine and Warfare, *Politics and the Life Sciences* ***13***, pp. 51-62.
23. Geissler, E. (ed.) (1986) *Biological and Toxin Weapons Today.* Oxford University Press, Oxford, pp. 135-137.
24. Ad Hoc Group of the States Parties to the BTWC (1997) *Rolling text of a Protocol to the Convention on the Prohibition of the Development, Production and Stockpiling of Bacteriological (Biological) and Toxin Weapons and on Their Destruction.* BWC/AD HOC GROUP/36, Annex I, 4 August.
25. Anonymus (1996) Libya and Iran seek ex Soviet scientists, *Science* **271**, p. 1485.
26. WEU (1955) Protokoll zur Änderung und Ergänzung des Brüsseler Vertrages und die Errichtung der Westeuropäischen Union unter Einschluß der Bundesrepublik Deutschland und Italiens vom 23.10.1954. *Bundesgesetzblatt* II, pp. 266-273.

CONVERSION OF CHEMICAL WEAPONS PRODUCTION FACILITIES UNDER THE CHEMICAL WEAPONS CONVENTION

DR THOMAS STOCK
Lobbe Xenex GmbH & Co., GERMANY

1. Introduction

The primary goal of the Chemical Weapons Convention (CWC) with respect to chemical weapon production facilities (CWPFs) is their destruction. However, there are provisions under the Convention which will allow conversion of CWPFs into: *(a)* facilities for destruction of chemical weapons (CW); and *(b)* facilities operating for purposes not prohibited by the CWC, such as commercial production.

The CWC has entered into force (EIF) 29 April 1997, after the 65th instrument of ratification has been deposited 31 October 1996 by Hungary. Today more than 100 countries have ratified or acceded to the CWC, among them the USA. Unfortunately, as already expected by many experts, the Russian Federation was not able to meet its commitment "to be among the original States Parties" at entry into force. Nevertheless, 31 October 1997, one year after Hungary had deposited its instrument of ratification, the Russian Duma gave approval to the CWC. Russia has deposited its instrument of ratification in November and took part in the December 1997 Conference of the States Parties as a full member of the CWC.

It seems to be necessary to remember who might be affected mostly by the issue "conversion of former CWPFs"? There is a clear answer, only states with CWPFs or former CWPFs (after 1 January 1946). The number of states being affected by the provisions for possible conversion of former CWPFs under the CWC will be rather limited.

After a long controversy in the negotiations at the Conference of Disarmament (CD) in Geneva, agreement was reached in 1988 that CWPFs must be destroyed. It was agreed that a conversion of CWPFs, in general, should be unacceptable. However, very limited conversion possibilities for CWPFs should be acceptable, focused on *(a)* the support of required CW destruction activities and *(b)* conducting not prohibited activities for peaceful purposes. Most of the conflicting discussions and debates during the CD negotiations where touched upon the differing views on destruction or conversion of CWPFs or former CWPFs from the than USSR and the US. The United States always underlined, based on its former "single-purpose CWPF situation", that it was more in favour of the general destruction concept, however, without objecting the conversion option in general. If conversion would be an option, so the US view, it has to be one under strong limitation. Irreversibility of converted CWPFs must be the overriding principle.

E. Geissler et al. (eds.), Conversion of Former BTW Facilities, 11-24.
© 1998 *Kluwer Academic Publishers. Printed in the Netherlands.*

In opposite, the USSR and later the Russian Federation always favoured a more moderate conversion definition. This position was based, among others, on the situation that former Soviet CWPFs were integrated into large civilian chemical production complexes and fitted into the same industrial infrastructure. Destruction or better "demolishing" these facilities would create serious difficulties for the other plants parts of such large complexes. In addition, Russia stated already in the late eighties, that some of the buildings, where production of CW had been taken places, had been converted into civilian commercial acting facilities. One example here is the former CWPF in Volgograd.

After intense bilateral negotiations a compromise was achieved which resulted in including in the draft treaty a permission of conversion of former CWPFs into facilities for legitimate activities. At this time, the late eighties, it seemed that both sides were satisfied with the results achieved so far in the treaty text. However, at the final end of negotiations in the Preparatory Commission at The Hague it was obvious that Russia was still not satisfied with the solution so far achieved.

In the time of transition from the Geneva negotiations to The Hague Preparatory Commission and the time before entry into force the expert discussion and deliberations didn't solve all the problems.

2. The CWC and CWPFs

The issue of conversion of former CWPFs is treated under the CWC in such a way, which clearly reflects the risk which CWPFs in general pose to the Convention. With other words, CWPFs are the facilities where chemical weapons have been manufactured in the past, before EIF. These facilities pose one of the highest risks to the purposes of the CWC.

This is reflected under Article I, paragraph 1, which states:

"Each State Party to this Convention undertakes never under any circumstances: *(c)* to [...], produce, [...], chemical weapons, [...]".

In addition, the same article outlines that

"Each State Party undertakes to destroy any chemical weapons production facilities it owns or possess, or that are located in any place under its jurisdiction or control, in accordance with the provisions of this Convention".

If any conversion of CWPFs by a State Party is envisaged, the CWC has to secure with its verification regime, that there will be no risk of any misuse of such facilities for prohibited purposes. This clearly is underlined by the fact, that conversion is not a right, is only an option under the CWC.

The emphasis under the CWC for CWPFs is on their destruction, because they are the facilities where CW had been produced or could still be produced (paragraph 4 of Article I). The destruction obligation under the CWC will be altered later by introducing the conversion option for the purpose to use the CWPF for destruction of CW or for purposes not prohibited by the Convention.

2.1 DEFINITION OF CWPFS

For the following discussion it seems to be important to remember what the definition of a CWPF contains. Here Article II of the CWC outlines in paragraph 8:

Chemical Weapons Production Facility:

(a) Means any equipment, as well as any building housing such equipment, that was designed, constructed or used at any time since 1 January 1946:

(i) As part of the stage in the production of chemicals ("final technological stage") where the material flows would contain, when the equipment is in operation:

(1) Any chemical listed in Schedule 1 in the Annex on Chemicals; or

(2) Any other chemical that has no use, above 1 tonne per year on the territory of a State Party or in any other place under the jurisdiction or control of a State Party, for purposes not prohibited under this Convention, but can be used for chemical weapons purposes;

or

(ii)For filling chemical weapons, including, inter alia, the filling of chemicals listed in Schedule 1 into munitions, devices or bulk storage containers; the filling of chemicals into containers that form part of assembled binary munitions and devices or into chemical submunitions that form part of assembled unitary munitions and devices, and the loading of the containers and chemical submunitions into the respective munitions and devices;

(b) Does not mean:

(i) Any facility having a production capacity for synthesis of chemicals specified in subparagraph (a) (i) that is less than 1 tonne;

(ii) Any facility in which a chemical specified in subparagraph (a) (i) is or was produced as an unavoidable by-product of activities for purposes not prohibited under this Convention, provided that the chemical does not exceed 3 per cent of the total product and that the facility is subject to declaration and inspection under the Annex on Implementation and Verification (hereinafter referred to as "Verification Annex"); or

(iii) The single small-scale facility for production of chemicals listed in Schedule 1 for purposes not prohibited under this Convention as referred to in Part VI of the Verification Annex.

In the definition of CWPFs, in contrast to chemical weapons, the primary focus is on the presence of Schedule 1 chemicals in the material flow of the equipment and construction including building housing them. The CWPFs are facilities manufacturing Schedule 1 chemicals or similar ones (with no legitimate uses above 1 tonne per annum per State Party but usable for CW purposes). The definition does not cover precursors or other toxic chemicals even if manufactured for weapons purposes (such as phosgene or hydrogen cyanide). Such facilities are covered under Article VI

(Activities not prohibited under this Convention), however, they are not subject to destruction under Article V (Chemical weapons production facilities).

The reference to the "final technological stage" has the consequence that facilities, part of the CWPF, but have produced some intermediates and functioning separate from the final technological stage, would not be covered by the definition. In addition, the construction of paragraph 8 takes care of the fact, that starting materials or intermediates are sometimes purchased or transferred from civilian industry. The final technological step only was conducted in the dedicated CWPF. In consequence, these facilities producing intermediates or starting materials don't have to be destroyed.

The problem of multipurpose plants is taken care in the definition under paragraph 8 by introducing the characteristic "designed, constructed or used" for manufacturing of Schedule 1 or similar chemicals, falling under the definition. Any other potential plant, especially with the capability to be used in the multipurpose mode, will be covered under the Verification Annex, Part IX, which is focused upon "other relevant facilities".

Under paragraph 8 there is also clear understanding that the CWC only copes with CWPFs operated after 1 January 1946. This cut-off date has practical implications for any state having had CWPFs in operation before and during World War II. Without such cut-off date, States Parties would be obliged to destroy plants, which are today after re-construction used as part of civilian chemical industry, but have been used as CWPF before 1946.

However, in the last 50 years not only Russia and the USA have had a CW production programme, so the obligations related to former CWPF might be of relevance for more States Parties. As the initial declarations after EIF submitted by several ratifiers have shown, there a few more countries with former CWPFs. By summer 1997 China, France, Japan, the UK and the USA have made Article V declarations.1 In addition, India's disclosure about its CW capability from 26 June 1997, should contain an information about an existing or former CWPF.2

In addition, the definition of CWPFs clearly indicates that under the scope of the definition also filling facilities and filling equipment, devices or bulk containers, assembly lines for unitary and binary as well as multicomponent CW, and loading lines for chemical submunitions will come too.

Last, but not least, former production facilities, being closed down, mothballed or converted after 1 January 1946, have to be declared and certain verification measures will be applied in the same manner as for still active CWPFs, always with the goal to identify and to verify that all provisions are correctly fulfilled.

The paragraph 8 under Article II contains few exclusions from the definition of CWPFs, as they are:

(a) precursor manufacturing facilities (also if they produce starting material for CW);

(b) small-scale production facilities (covered under Article VI);

(c) the single small-scale facility (covered under Article VI);

---

1 '28 July', *The CBW Convention Bulletin*, issue no. 37. Sep. 1997, p. 31.

2 '26 June', *The CBW Convention Bulletin*, issue no. 37. Sep. 1997, pp. 24-25.

(d) facilities producing Schedule 1 chemicals as an unavoidable by-product, below 3 per cent concentration;

In summing up, the difficulty under the CWC to define CWPFs by taking care of the general purpose criteria as defined under Article I, was to establish a framework which allows to distinguish between production facilities in general and facilities which are specifically designed in their last technological step to produce CW.

3. Conversion of CWPFs under the CWC

Under Article V clear guidance is give to the scope of any planned conversion of CWPFs. Paragraphs 12 and 13 provide for two different options of conversion of a CWPF: *(a)* temporary conversion into chemical weapon destruction facility (CWDF);3 *(b)* final conversion into a facility for purposes not prohibited. In addition, the CWC does distinguish between a conversion which had been conducted already before EIF and a conversion which is anticipated after EIF.

3.1 TEMPORARY CONVERSION INTO CHEMICAL WEAPON DESTRUCTION FACILITY

Paragraph 12 of Article V provides the guidelines for temporary conversion into CWDFs as the following:

Chemical weapons production facilities specified in paragraph 1 may be temporarily converted for destruction of chemical weapons in accordance with Part V, paragraphs 18 to 25, of the Verification Annex. Such a converted facility must be destroyed as soon as it is no longer in use for destruction of chemical weapons but, in any case, not later than 10 years after entry into force of this Convention.

For the verification and the other provisions related to the temporarily converted CWPF the Verification Annex, Part V, paragraphs 18 to 25 provide guidance.

The basis provisions here are:

(a) CWDFs, temporarily converted from CWPFs, remain original CWPFs; (paragraph 23);

(b) CWDFs, converted from CWPFs, have to be destroyed not later than 10 years after EIF (paragraph 23);

(c) converted facilities for CW destruction shall not be more fit to resume CW production than a CWPF which has been closed and is under maintenance (paragraph 22);

(d) the set of measures carried out for the purpose of conversion shall not be less than that which is provided for disabling of over CWPFs (paragraph 25);

(e) measures for conversion of CWPFs into CWDFs shall be facility specific (paragraph 24);

---

3 Temporary has here the meaning during the 10-years destruction period for CWPFs.

(f) CWPFs, converted into CWDFs before EIF, have to be declared under the category of former CWPFs; initial visits by OPCW inspectors have to confirm that conversion has been conducted in accordance with the Convention principles and requirements (paragraph 19);

(g) if a State Party intends to convert a temporarily a CWPFs into a CWDF after EIF, a general conversion plan for such a facility has to be submitted to the Technical Secretariat; 180 days remain for arranging all necessary details of the planned conversion between the State Party and the Technical Secretariat, after the plan has been submitted (paragraphs 20 and 21);

(h) annual plans on the destruction of CW (see Verification Annex Part IV (A)) have to be submitted, as long as the converted CWPF is being used as a CWDF (paragraph 20); and

(i) all the operations in the temporarily converted CWDF, originally being a CWPF, will be subject to the verification measures as set forth for systematic on-site verification at CWDF (Verification Annex, Part IV(A), paragraphs 65-70).

Paragraph 18 sets forth the overriding principle for the temporarily conversion by "the regime for the temporarily converted facility is at least as stringent as the regime for chemical weapons production facilities that have not been converted". With other words, any temporarily converted CWPF, used as an CWDF, will not ease the verification regime.

Last, but not least, after the temporarily converted CWPF has ceased destruction operations, it will be handed over to the provisions for destruction of CW (as outlined under the Verification Annex, Part V, paragraphs 32-33).

3.2 FINAL CONVERSION INTO A FACILITY FOR PURPOSES NOT PROHIBITED

In the case of conversion of CWPFs into facilities for purposes not prohibited under the CWC paragraph 13 outlines:

A State Party may request, in exceptional cases of compelling need, permission to use a chemical weapons production facility specified in paragraph 1 for purposes not prohibited under this Convention. Upon the recommendation of the Executive Council, the Conference of the States Parties shall decide whether or not to approve the request and shall establish the conditions upon which approval is contingent in accordance with Part V, Section D, of the Verification Annex.

Of major importance in the above paragraph is the understanding of the phrase "compelling need", which has to be defined by the State Party through its request for conversion to the OPCW. Such a request has to be approved by the Conference of States Parties after recommendation of the Executive Council. There are no guidelines to determine the compelling need.

In addition, Article V outlines the goal for any conversion for purposes not prohibited in paragraph 14:

The chemical weapons production facility shall be converted in such a manner that the converted facility is not more capable of being reconverted into a chemical weapons production facility than any other facility used for industrial, agricultural, research, medical, pharmaceutical or other peaceful purposes not involving chemicals listed in Schedule 1.

The procedures for requesting a conversion into facilities for not prohibited activities, the conditions and verification provisions are set forth under the verification Annex V, Part D, paragraphs 64-86.

The procedures for *filing the request* contain the following main provisions:

(a) request can be submitted for each CWPF, including those already converted before EIF; in other words, every former CWPF, used after 1 January 1946, for which no request has been mad, will remain a CWPF and must be declared and destroyed accordingly to Article V (paragraph 64);

(b) for CWPFs which are being used for purposes not prohibited under the CWC at EIF, the request containing detailed information on facility and conversion, should be made 30 days after EIF for the State Party (paragraph 65);

(c) for CWPFs for which conversion into facilities for purposes not prohibited under the CWC is planned after EIF, the State Party has to submit the request not later than 30 days after the decision to convert; the request should contain details on facility and planned conversion; in addition, the request must contain the details for the economic needs of such a conversion (paragraph 66);

(d) any request for conversion of CWPFs into facilities for not prohibited purposes has to be filed not later than four years after EIF of the CWC for the State Party (paragraph 66); and

(e) a State Party already converted a CWPF or requesting conversion into facilities for not prohibited purposes might propose additional measures to build confidence, such as additional verification measures (paragraph 67);

After filing the request for conversion a CWPF into a facility for not prohibited purposes by a State Party there might be a time before the final decision will be taken by the Conference of the States Parties. During this time different actions are foreseen, as outlined under paragraph 68 and 69 of the Verification Annex, Part V. Among them, the State Party has to certify, that the already converted facility does not use specialized equipment or buildings, or that they have been rendered inactive.

Most important are the *conditions for conversion* as outlined under paragraphs 70-72:

- As a condition for conversion of a chemical weapons production facility for purposes not prohibited under this Convention, all specialized equipment at the facility must be destroyed and all special features of buildings and structures that distinguish them from buildings and structures normally used for purposes not prohibited under this Convention and not involving Schedule 1 chemicals must be eliminated.

- A converted facility shall not be used:

(a) For any activity involving production, processing, or consumption of a Schedule 1 chemical or a Schedule 2 chemical; or

(b) For the production of any highly toxic chemical, including any highly toxic organophosphorus chemical, or for any other activity that would require special equipment for handling highly toxic or highly corrosive chemicals, unless the Executive Council decides that such production or activity would pose no risk to the object and purpose of this Convention, taking into account criteria for toxicity, corrosiveness and, if applicable, other technical factors, to be considered and approved by the Conference pursuant to Article VIII, paragraph 21 (i).

- Conversion of a chemical weapons production facility shall be completed not later than six years after entry into force of this Convention.

Taking the time-limit for requesting any conversion from paragraph 66, not later than four years after the CWC enters into force for the State Party, and the time-limit for completion of any conversion, there might be very limited flexibility, if a State Party requests conversion at the very last moment. In addition, if a State Party would accedes to the CWC after EIF of the Convention very late, the conversion option might be limited with respect to the 6 years window.

The Convention provides provisions for the *decision making* upon any State Party's request about conversion by the Executive Council and the Conference of the States Parties (paragraphs 73-76). In detail the process contains the following elements:

(a) conducting an initial inspection of the facility, for which conversion is requested, not later than 90 days of request is filed; the purpose of such inspection is to determine the accuracy of the provided information, to obtain information about proposed converted facility and assess the conditions for the use (paragraph 73);

(b) based upon the initial visit, the Director-General submits a report, containing recommendations, to the Executive Council, the Conference of the States Parties and all States Parties (paragraph 73);

(c) in case the facility has been converted before EIF and the conditions for conversion are not met, the Director-General informs the Executive Council, which may required *(1)* implementation of the measures, or *(2)* shut down of the facility and removal of specialized equipment (paragraph 74);

(d) the decision making process in the Conference of the States Parties is based upon a decision on a "matter of substance" (two-thirds majority is required); if individual States Parties are vetoing, there might be a possibility to call in a 90-days consultation period (paragraph 75); and

(e) after the request of a State Party about conversion is approved a facility agreement, containing all established conditions for the conversion, has to be concluded, the agreement should contain the detailed verification provisions and any limitations on conversion (paragraph 76).

However, the decision making process might take some time and there is no assurance that the original request by an individual State Party will receive full approval, without additional recommendations by the Conference of the States Parties.

In case, that a State Party will request a conversion of a CWPF when the CWC enters into force for it, there are specific provisions for submission the detailed plan for conversion (paragraphs 77 and 78). The plan has normally to be submitted 180 days before conversion is anticipated to start, however, taking the above outlined decision making process into consideration, the plan might even be submitted much earlier. (The plan should contain many technical details.)

Last, but not least, the Convention provides for a mechanism to review the detailed plans for conversion (paragraphs 79-84).

There are two additional provisions under the CWC which are of relevance for the conversion of former CWPFs into facilities for not prohibited purposes.

For the first ten years after completion of conversion the basis requirements for the *verification regime* are established (see paragraph 85). This regime is based upon:

(a) the unimpeded access to the facility at any time;
(b) inspectors have right to observe all areas, all activities and all items of equipment;
(c) inspectors have the right to receive samples from any area of the facility and to analyse them; and
(d) inspectors shall have the right to managed access to the plant site at which the facility is located.

During the 10 year period, the State Party has to report on the activities at the converted facility annually. After this period, the Executive Council, upon the Technical Secretariat's recommendations, has to decide on the nature of continued verification measures.

Most important, as outlined under paragraph 86, the State Party has to meet the costs of verification, if not decided otherwise by the Executive Council. However, there are possibilities to reduce these direct inspection costs by complementary mechanisms (Article V, paragraph 17, Article VIII, paragraph 7). Bilateral or multilateral agreements on destruction, if in place, could reduce the verification activities of the OPCW. Based upon the concept of proportionality, the States Parties, members in such agreements, could reduce their direct inspection costs.

In summing up, any anticipated conversion of a CWPF into a facility for not prohibited purposes will bring up an additional number of obligations for a State Party, which intends to do so. Based upon the general attitude to conversion "there is no right for conversions, there is only an option", the State Party should keep in mind all *pros* and *cons* for such a step. The major stumbling block might be the nature of the decision making process in the Conference of the States Parties, which will internationalize the process without having the assurance about a positive decision if a request is filed from the beginning. In addition, extra costs will be allocated to the State Party which intends to go for conversion.

4. CWPFs conversion or destruction?

Any consideration about conversion of CWPFs *(a)* temporarily into CWDFs, or *(b)* into facilities for not prohibited purposes, has to take into account the anticipated verification requirements. In addition, based upon the requirement to borne the costs of verification of converted facilities, the State Party has to decide upon the size of the additional economic burden. The overriding question for a State Party having a CWPF, existing at EIF, is to consider the pros and cons of an anticipated conversion or destruction.

4.1 DESTRUCTION OF CWPFS

The order of destruction of CWPFs is based on the principle of "levelling out".4 The CWC here defines three specific time intervals for facilities producing Schedule 1 chemicals: *(1)* years 2-5; *(2)* years 6-8 and *(3)* years 9-10 after EIF of CWC (as specified under Verification Annex, IV Part V paragraph 30). The destruction has to start not later than one year after EIF for the State Party. Based upon these time intervals the individual State Party is required to plan and to administer the destruction in detail. However, the CWC provides some guidance for conducting the process:

- production capacity shall be used as a comparison factor between the facilities (based upon the rule for binary chemical weapons the capacity shall be expressed in agent tonnes);
- appropriate agreed levels of production capacity shall be established for the end of the eight year after EIF;
- production capacity that exceeds the relevant level shall be destroyed in equal increments during the first two destruction periods; (2-5 years, 6-8 years);
- a requirement to destroy a given amount of capacity shall entail a requirement to destroy any other CWPF that supplied the Schedule 1 facility or filled the Schedule 1 chemicals produced there into munitions or devices; and
- CWPFs that have been converted temporarily into CWDFs shall continue to be subject to the obligation to destroy capacity according the provisions applied for normal CWDF.

For CWPFs not producing Schedule 1 chemicals, the destruction has to be started not later than one year after EIF of CWC for the State Party and has to be completed

---

4 Also under the order of destruction of CW the same principle has been applied. "Levelling out" is based upon the concept that with unequal starting positions for different States Parties, destruction of capacity shall be performed in such a way that over a certain time period these inequalities would be diminished until the final phase, after year 8, there all States Parties are obliged to destroy CWPFs at the same pace. (this is a concept developed during the US-Soviet Bilaterals).

not later than 5 years also after EIF of CWC, as specified under Verification Annex, IV Part V paragraph 31.

Last, but not least, the CWC does not contain any requirement which will preclude a State Party from destroying CWPFs at a faster pace as required for Schedule 1 chemical producing facilities and other facilities.

Taking this time-schedule into account, it is more than obvious, that for a State Party acceding late after EIF, the order and rate of destruction may be significantly compressed. Here the CWC outlines under paragraph 30 of Verification Annex, IV Part V: 'For States which become a Party after entry into force of this Convention, the destruction periods shall be adapted, taking into account paragraphs 28 and 29".

The main point under paragraph 28 is that the order of destruction has to be based on the principle of levelling out, besides general points, such as undiminished security during the destruction period, confidence-building in the early part of the destruction stage, etc. Based upon this levelling out principle late-comers will be faced with the difficulty to destroy with much faster pace to arrive at the 8th year residual level, which has been agreed among the original ratifiers of the CWC.

4.2 CONVERSION OF CWPFS INTO FACILITIES TO NON-PROHIBITED USES

As outlined under Article V a State Party my request approval to the conversion of a CWPF into a facility for purposes not prohibited by CWC. The most important point here, the State Party must show that there is a "compelling need" for doing this conversion. The Executive Council has to give recommendation to the Conference of the States Parties, which has to decide about the request.

The detailed procedures for the conversion of CWPFs to purposes not prohibited under the CWC have been discussed above.

However, special factors following directly from the Convention, will influence any decision to go for conversion into facilities used for purposes not prohibited. Such factors are:

(a) conversion of CWPFs is a possibility, not a right (Article V, paragraph 13);

(b) additional information and declaration is required in case of conversion is planned (Verification Annex Part V, Section D, paragraphs 65-67; 77-78);

(c) converted CWPFs will be subject for more stringent verification (Verification Annex Part V, Section D, paragraph 85);

(d) cost of verification have to be borne by the State Party (Article V, paragraph 19, Verification Annex Part V, Section D, paragraph 86); and

(e) conversion of CWPFs has to be conducted under time constraints (Verification Annex Part V, Section D, paragraphs 65-66, 69, 75-76, 77-78, 80 and 85).

4.3 TEMPORARILY CONVERSION OF CWPFS INTO CWDFS

CWPFs may be converted temporarily into CWDFs. The verification provisions are less rigorous than for conversion into facilities used for purposes not prohibited. Of major importance, only the Technical Secretariat and the State Party need to agree on details of conversion. In addition, there are no specific intermediate time limitations for temporarily conversion into CWDFs (the only one: converted CWPFs into CWDFs must be destroyed not later than ten years after CWC has entered into force).

5. Conclusions

With EIF of the CWC in late April 1997 a clear political signal has been given by many ratifying countries to continue and to complete the process of chemical disarmament.

Every CWPF has to be destroyed, however, the CWC envisages two exceptions: *(a)* temporarily conversion into CWDF; *(b)* conversion into a facility which will conduct activities not prohibited by the Convention.

The conversion of former CWPFs into CWDFs or into facilities used for purposes not prohibited is an option which might be relevant only few States Parties only.

The temporarily conversion into a CWDF is limited with respect to time-frame, because the destruction of CW has to be finished ten years after EIF of the CWC. After these ten years such a converted CWDF has to cease destruction, the same procedures for closing down the facility and immediate destruction, as for normal CWPFs, will be applied. This clearly indicates that the CWC treats CWDFs, converted from former CWPFs, still as CWPFs. There is no excuses, they have to be destroyed not later than 10 years after EIF of the CWC, not EIF for the State Party of concern.

The conversion into facilities for purposes not prohibited has to be seen as an approach having had developed from the negotiation history, especially the impact which the two former super-powers had on the negotiations, and the understanding that some facility-parts might have an commercial chemical production relevance. However, the reality after EIF will provide evidence how realistic this approach might be (because of the cut-off date, 1 January 1946, the conversion of former production facilities into facilities which do have only commercial application might be seen in a much broader sense).

After detailed analysis of all CWC requirements relevant for any planned conversion of former CWPFs there seems to be need for realistic consideration of the *pros* and *cons*. Any delay in a possible ratification of CWC by an individual State Party will compress some of the anticipated time-schedules for destruction in general and conversion in particular. Despite the additional economic burden, any envisaged conversion of former CWPFs into facilities for purposes not prohibited will definitely increase the costs by adding additional burden on intense declaration and inspection requirements.

Demanding conversion of CWPFs into facilities for not prohibited purposes, there is now direct possibility under the CWC. Asking for conversion is the only way. However, the Conference of the States Parties as the whole is the only organ who decides on such a request.

If asking about the applicability of the CWC approach to conversion for the Biological Weapons Convention (BWC), one major difference has to be kept in mind. CWPFs are quite different compared to any possible BW production facilities, especially in size and technological set-up. In addition, the "dual-use capability" of equipment is of different magnitude. Production equipment for BW agents need not differ from that used in a pharmaceutical industry. The technology and equipment to do research, development and production of biological agents can be identical with which is used in medical research.

However, taking the CWC as an pragmatic point of departure for any possible conversion approaches under the BWC the following lessons can be learnt:

- BW production facilities have to be defined via equipment, installations, buildings, etc.;
- BW production facilities have to be declared based upon clear guidelines;
- BW production facilities have to be initially inspected to verify the declarations;
- past BW production facilities, already converted, should be declared and subject to initial visits;
- any anticipated conversion of a dedicated BW production facility has to be requested by the State Party in advance, a conversion plan has to be submitted;
- BW production facilities under conversion should receive regular inspections;
- for any BW production facility under conversion there should be a facility agreement to guide the inspection obligations; and
- any converted BW production facility should submit annually a simple declaration.

There is clear evidence from the conducted analysis that BW production facility conversion is only an option. However, without clear definition of "what is a BW production facility and which compartments are included" it will be more than difficult to proceed. Strong verification should be required, if conversion will be conducted. Extra economic burden and verification costs have to be included in watching the *pros* and *cons* of any anticipated conversion.

Annex I: Conversion of CWPFs under the CWC

State Party's obligations to submit information to the OPCW and particular actions, if conversion it being planned (at EIF) or had been succeeded (already before EIF):

A: Conversion of CWPFs into CWDFs

(1) 30 days after EIF/or 30 days after decision is taken:
statement about intended conversion into CWDF
(2) 30 days after EIF/or 30 days after decision is taken:
initial conversion plan envisaged time-frame for conversion and use
(3) 90 days after EIF:
all CWPFs must be closed
(4) 90 days after EIF/or after ratification:
inspectors have the right to visit the plant
(5) 120 days after EIF:
latest date for familiarization visit of TS to any former CWPF which is planned to be converted into CWDF
(6) 150 days after EIF/or 120 days after notification:
conclude transition agreement considering additional inspections and measures
(7) 180 days after EIF:
transition agreement might be approved and conversion can start
(8) 10 years after EIF:
CWDF has to be closed

B: Conversion of CWPFs into facilities for non prohibited purposes

(1) 30 days after EIF/30 days after decision is taken:
submit a request on planned conversion
(2) not later than 90 days after EIF:
close any CWPF
(3) not later than 90 days after request on conversion has been submitted:
initial inspection of TS
(4) not later than 120 days after EIF:
latest date for submission of detailed plans for the planned conversion of CWPF into facility for non prohibited purposes
(5) 180 days + n.n. days:
Conference of the States Parties has to consider the conversion request for CWPFs into facilities for non prohibited purposes
(6) less than 6 years after EIF:
conversion must be completed
(7) 1-10 years after EIF:
unlimited access to converted facility
annual reporting on activities
(8) > 10 years:
new procedures for verification

BIOTECHNOLOGY IN A PEACEFUL WORLD ECONOMY

MALCOLM R. DANDO CBiol MIBiol
Professor of International Security
Department of Peace Studies
University of Bradford
Bradford BD7 1DP
West Yorkshire UK

1. Introduction

At the Fourth Review Conference of the Biological and Toxin Weapons Convention, in late 1996, the head of the US delegation noted that "twice as many countries now have or are actively pursuing offensive biological weapons capabilities as when the Convention went into force".[1] It has always been difficult to be sure which countries are involved. In 1993 the US Office of Technology Assessment, for example, had to list "Countries Generally Reported as Having Undeclared Offensive Biological Warfare Programs" and then named Iran, Iraq, Israel, Libya and Syria in the Middle East and China, North Korea and Taiwan in East Asia.[2] Whilst information on some countries has become more specific in recent official statements[3] the scale of the problem in most countries is still a mystery to the general public.

There are two exceptions to this general rule: Iraq, where UNSCOM has uncovered a great deal of information on the pre-1991 offensive programme,[4] and the former Soviet Union. The facilities used by Iraq are not of great concern here as long as they remain under UN supervision. The facilities used in the former Soviet Union are, however, of central concern for any review of the possibility of conversion. The available public information leaves no doubt that this was a large-scale programme involving many different facilities over a protracted period of time.[5]

The conversion of any of these facilities to peaceful production at reasonable cost is undoubtedly to be welcomed.[6,7] Yet despite the fact that some of them would appear technically to have been ideal candidates,[8] it is undoubtedly true that effective economical conversion of individual defence facilities is very difficult to achieve. The difficulties are not primarily technical but, in moving from dealing with a single Ministry of Defence customer to a rapidly changing civil market with multiple customers and competitors, Western experience strongly suggests that conversion can only really be effective if carried out within some wider context such as the regeneration of an entire region.[9] It is not greatly surprising, therefore, to find pessimism about the Russian experience of conversion to date.[10]

The successful development of the Russian economy is, of course, crucial for European and world security in the next century.[11] Also, given the probable

E. Geissler et al. (eds.), Conversion of Former BTW Facilities, 25-43.
© 1998 *Kluwer Academic Publishers. Printed in the Netherlands.*

contribution that modern biotechnology will make to future industrial growth,[12] the success of civil biotechnology in Russia is a specific issue of considerable importance in itself as well as an indicator of the potential scale of the problem and the possible rewards for successful conversion in other countries.

My task here, however, is to discuss the context within which conversion might succeed rather than the problem of facility conversions in particular countries. Looked at in a broad sense, the problem is that through this war-ravaged century major nations have found it necessary to build up and maintain what are, by historical standards, huge war industries. Thus the widespread disappointment that a promised 'peace dividend' did not result from the ending of the Cold War is quite misplaced. It was to be expected that, given prevailing realist beliefs, military industry would be reorientated to the new products made possible as the information technology revolution[13] continued to impact on military affairs and that 'savings' would be brought about by redundancies and downsizing of older elements in the various defence industrial bases.[14] The fundamental question for effective conversion, therefore, is whether means can be found to turn our advanced military economies into advanced civil economies.

In order to deal with that question I wish first to discuss an older and broader view of conversion than that centred on facility conversion: that is, the idea of a 'peaceful economy'. Within that context I will then look briefly at some of the tasks that biotechnology might usefully address in a future peaceful world economy. In the third section of the paper I shall attempt to review the capabilities, and apparent intentions, of the industry in relation to these tasks. The paper ends with a consideration of what might be done to correct any mismatch between the ideals set out and the present performance of the industry, particularly as we consider specific examples of industrial facility conversion.

2. The Idea of a Peaceful Economy

Whilst the research group to which I belong has contributed to the debate on conversion since the Cold War ended,[9,14] it was actually much more active in this field of research during the era's latter stages.[15,16] Given that we work within a Department of Peace Studies, it is not surprising that ideas on conversion were influenced by other scholars who saw the war-producing economy as but an aspect of modern industrial society's exploitation of the natural world and human beings in the reckless pursuit of material abundance.[17]

TABLE 1A. Characteristics of an economy of violence *

- Large-scale production, hierarchically and bureaucratically controlled;
- High levels of conflict over possession of the world's resources, andover distribution of rewards for production;
- A complex of military interests in industry and the economy, and the domination of the economic life of towns, localities, regions, and national economies by defence production and military research and development;
- Mass-production, standardisation, and the destruction of skills and autonomy;
- A high demand for energy to fuel the industrial machine, and pollution and exploitation of the environment.

* From reference 18

From this standpoint, Tom Woodhouse, an historian who studied the British Labour movement and who led our work on conversion during the 1980s, called on Lewis Mumford's analysis of the difference between 'authoritarian technics' and 'democratic technics'[18] to differentiate between an economy of violence (Table 1A) and a peaceful economy (Table 1B). Lest these ideals seem far-fetched, it should be remembered that they informed the creation, for example, of the Scott Bader Commonwealth (Table 2) which is now a £100 million-a-year chemical resins company, and that some current ideas on a 'stakeholder' society appear to derive from similar sources.[19]

In the United States, Mumford's analysis was also influential, as was Seymour Melman's critique of the way that a permanent war economy became established following the Second World War.[20] In 1988, in a special edition of the *Bulletin of Peace Proposals* on 'Swords into Ploughshares: Conversion of Military Production to Socially Useful Purposes', Michael Renner of Columbia University drew on these American analyses to detail the differences between a 'conflict economy' and a 'peaceful economy'.[21] These differences were similar to those set out by the British authors in Tables 1A and B but were illuminated by the questions posed in Table 3.

There is much that is prophetic in Renner's article about the lost opportunities - in both East and West - following the ending of the Cold War. The message that needs emphasising, it seems to me, is encapsulated in the following passage:

"...Economic conversion, most immediately, is concerned with the changeover from military to civilian economy. However, in a broader sense, it is also concerned with the kind of civilian economy to emerge. For to prevent a return to using military means to 'settle' conflicts, it is advisable to prevent those conflicts that arise from fundamental inequalities..."

Therefore, in Renner's view:

"...Economic conversion, then, involves more than just a move away from all things military. It is an essential tool to restructure an economy, to create a more just, equitable, and sustainable society."

Achieving economic conversion may involve a long and difficult road for human society but without that goal, the maintenance of huge war economies is almost inevitable. We must therefore keep the goal clearly in mind when considering conversion of biotechnology facilities formerly used in offensive biological warfare programmes.

TABLE 1B. Characteristics of a peaceful economy *

- Democratic control and co-operative working patterns;
-A diversity of modes of economic activity, placing primary emphasis on human-centred organisation and on meaningful work;
- Production of arms forming a small part of economic activity;
- Security based on a conception of defence far wider than a dependence on technology (weapons);
- Low levels of violence within and between societies, and the availability of skills and resources to constructively resolve the conflicts which do occur;
- A regard for finite resources and the delicate balance of the environment;
- An awareness of the global consequences of economic activity.

* From reference 18

TABLE 2. Key principles of the Scott Bader Commonwealth (1951) *

- The firm should not grow beyond a limited size;
- There should be a fixed ratio between the highest and lowest paid;
- The members of the Commonwealth are partners and not employees;
- The Board of Directors should be accountable to the Commonwealth;
- The net profits should be divided between the company and the Commonwealth, 60% to the former, and 40% to the latter;
- Half of the profits appropriated by the Commonwealth to be paid in bonuses to the members, and half to charitable purposes outside the organisation.

* From reference 18

TABLE 3. Questions illuminating the differences between a peaceful and a war economy *

Economic Development/Rationality
- Who controls the flow and distribution of wealth ?
- Is the economy based on exploitative relationships ?

Production Process/Work Organisation
- Is the decision-making process collective and participatory, or hierarchical ?

Choice of Technology/Product Choice
- What technology/materials are used ?
- To what extent do they contribute to pollution, hazardous work conditions, large-scale production units, and hierarchical control ?

Capital Intensity
- To what extent does the choice of technology increase or decrease capital intensity?
- To what extent does the choice of technology affect an economy's dependence on imported capital equipment ?

Scale/Complexity
- To what extent does technology 'necessitate', or contribute to, the large scale and complexity of productive units ?

World Market (Inter-) Dependence
- To what extent is the local/regional economy being linked to centres of the global economy ? How much local control is there over the nature of that link?
- Is this link of a beneficial nature or is it a burden on the local economy ?

Separation of Production and Consumption
- To what extent are individuals and communities separated into producers and consumers ?

Social Regulation
- Do the economic structures imply the creation of bureaucracies to implement and supervise the use of technologies ?
- To what extent is formal rationality substituted for substantive moral decision- making?

Resource Use/Ecological Impact
- To what extent does the turnover-time of production and consumption of goods and services increase faster than the turnover-time of the relevant ecosystem ?
- What is the impact on the local, regional and global ecosystem ?
- How are the goods of the production system tied to the ecological system ?

*From reference 21

3. Some Tasks for Biotechnology in a Peaceful World Economy

In light of the objectives embodied in the idea of a peaceful world economy the principle problem, as we rush towards the endpoint of globalisation,[22] is the enormous gulf between the rich and poor countries of the world today. As Erskine Childers pointed out in his introduction to a recent collection of essays on the challenges facing the United Nations,[23] the richest fifth of the world's population has 82.7 per cent of the world's GNP whilst the poorest fifth has just 1.4 per cent (Table 4). Moreover, the population of the developing world continues to grow much faster than that of the developed world, thereby accentuating these differences.

If biotechnology is to be one of the key industries of the next century[12] then there are clearly at least two tasks to which it might be applied to help the developing world. Firstly, although *The Economist*, in the lead-up to the 1996 World Food Summit in Rome, argued that:[24]
"...price rises and destocking are not early signs that world food production is nearing some natural limit....In the longer run, food output has risen faster than population in the past 30 years....All that provides reason to downplay Malthusian worries about food and hunger..."
the situation, when examined in detail, is considerably more worrying.

TABLE 4. Global economic disparities: distribution of economic activity 1989 (percentages) *

	Richest Fifth	Poorest Fifth
GNP	82.7	1.4
World trade	81.2	1.0
Commercial lending	94.6	0.2
Domestic savings	80.6	1.0
Domestic investment	80.5	1.3

* From reference 23

In considering the strategic issues raised by food supply, Leif Rosenberger[25] noted that it is wrong to concentrate on commercial demand for food as many economists would be inclined to do. Rather, he suggests, we must pay attention to 'social demand':

"...Social demand identifies the quantity of food that the world needs to feed all of its people, irrespective of commercial demand for grain in the marketplace...And if we can think of the world food supply in such a holistic context, we discover that the social demand for food is increasing..."

There are not only more people to feed, but as countries such as China develop economically their people demand more meat in their diet - and that is a very demanding and inefficient use of grain production. One pound of beef takes eight to ten pounds of grain to produce whereas:

"...that same eight pounds of grain can supply about ten times as many calories and more than four times as much protein when consumed directly by humans."

At present growth rates the world's population of some 6 billion is set to double by 2035, so meeting the demand for food will be no easy task.

Optimists argue that similar pessimistic conclusions were drawn 30 years ago,[24] but that the 'Green Revolution' brought an amazing and quite unexpected leap in agricultural productivity in the developing world. It is well known that Norman Borlaug:[25]

"...whose ingenuity and innovation combined modern, higher yielding seed varieties of rice, wheat, and maize with intensive and innovative use of inputs such as fertilizers, irrigation, and pesticides..."

was responsible in the main for this development. It is also clear that the Green Revolution has yet to be fully implemented, for example in Africa, and that if it can be fully implemented some further gains are possible.

Yet many experts ask whether that will be sufficient to cope with the increasing demand, particularly as the rate of growth in productivity seems to be slowing down. As Rosenberger noted:[25]

"...From 1961 to 1992, growth in world agricultural production decelerated, dropping from three percent annually in the 1960s to 2.3 percent per year in the 1970s and two percent during the 1980-1992 period.

Given these declining rates, can the capacity of the world agricultural system continue to feed a world population increasing by about 90 million people each year?"

He argues that the outcome is very uncertain and that a broad approach using a variety of policy options - expanding land use, irrigation, fertilisers and the determined application of modern biotechnology - will be required. This broad approach was endorsed strongly by the head of the UK Institute of Arable Crops Research at Rothamsted. The *Financial Times*[26] reported him urging that nothing be ruled out and quoted him as saying:

"...We don't yet know how to change the productivity of our major cereal crops..."

And whilst it could take 10 to 20 years of research to find out, he continued:

"...We can't sit back and say that all the other things like storage, distribution and politics will come right, so we don't need to tackle basic plant productivity. It could be a bit late by then to find out whether the optimists or pessimists were right."

In short, there is a real need to apply modern biotechnology to the task of feeding the world in the next century *now*.

The second task to which biotechnology could undoubtedly contribute is, of course, the improvement of the health of people in the developing world. As the preface to the 1993 World Bank report, *Disease Control Priorities in Developing Countries*, noted:[27]

"Between 1950 and 1960, life expectancy in developing countries increased from forty to sixty-three years with a concomitant rise in the incidence of the noncommunicable diseases of adults and the elderly..."

But:

"...there remains a huge unfinished agenda for dealing with undernutrition and the communicable childhood diseases..."

Estimates of the deaths caused in 1985 by infectious and parasitic diseases in developing countries are set out in Table 5.

TABLE 5. Estimated causes of death in developing countries, by age, 1985 (thousands) *

Cause	Age		
	Under five	Five and over	All ages
Infectious and parasitic diseases	10,500	6,500	17,000
Diarrhoeal diseases	4,000	1,000	5,000
Tuberculosis	300	2,700	3,000
Acute respiratory diseases	4,300	2,000	6,300
Measles, whooping cough, and diphtheria	1,500	-	1,500
Other acute respiratory diseases	2,800	2,000	4,800
Other measles and whooping cough[a]	700	-	700
Malaria	750	250	1,000
Schistosomiasis	-	200	200
Other infectious and parasitic diseases	450	350	800[b]

[a] Does not include 400,000 measles-related deaths included under diarrhoeal diseases
[b] Some deaths may be attributed to malaria
* From reference 27

The report concluded that:

"...Taken together, the vaccine-preventable diseases either were the cause of, or were closely associated with, 3.7 million deaths of young children in 1985, of which 2 million were from measles, 600,000 from whooping cough, 300,000 from tuberculosis, about 800,000 from neonatal tetanus, and 25,000 from poliomyelitis."

This represents a terrible burden of unnecessary suffering, when vaccination is a very cost-effective method of treatment.[27]

Yet despite the international community's success in eradicating smallpox and the ongoing campaign to eliminate polio,[28] currently available vaccines for childhood diseases are far from adequate. The early 1990s vaccination schedule of the World Health Organisation's Expanded Programme for Immunization (EPI)[29] is set out in Table 6. In 1974, when the programme began, there was only five per cent coverage globally. By 1991 that had reached 80 per cent. However, further improvement in coverage is thought to be increasingly difficult, partly because deficiencies in the available vaccines make delivery costs high.

TABLE 6. The World Health Organisation's Expanded Programme for Immunisation (EPI) *

Time	Vaccine
Shortly after birth	Tuberculosis (bacillus Calmette-Guérin, BCG)
2, 6 and 12 weeks	Diptheria-pertussis-tetanus (DPT)
Birth, and with DPT inoculations	Oral polio vaccine (OPV)
Between 9 and 12 months	Measles vaccine
Prior to delivery	Minimum two doses of tetanus toxoid to mother

* From reference 29

From the perspective of the public health official and vaccination programme manager the deficiencies of the available vaccines include:[29]

"...the requirement for multiple doses to achieve the full protective effect, the requirement for subcutaneous injections, and the need for an uninterrupted cold chain."

Thus:

"...It is apparent that improved vaccines containing multiple antigens, given orally and requiring fewer doses, would be an immense benefit to the vaccination programs. Increased efficiency would improve vaccination coverage and reduce delivery costs. Heat-stable vaccines would also be a great benefit in reducing losses and the logistic burden."

Again, it would seem probable that modern biotechnology could be effectively applied to dealing with vaccine deficiencies as a major part of an integrated programme.

Putting modern biotechnology in the *developed* world to the task of quickly and effectively dealing with the problem of infectious diseases in the developing world would not just be an act of charity. As the recent report from the US Institute of Medicine,[30] *America's Vital Interest in Global Health*, strongly stressed, in a world where one million people travel between the developing and industrial worlds each week the danger of the spread of emerging and re-emerging diseases[31,32] is ever present. A similar argument for self-interest can be made regarding food supply, for if the developed world chooses to ignore famine in the developing world it will inevitably have to deal with the conflict, mass migrations and general instability that is likely to result.

4. Biotechnology Today

It seems reasonable, when considering the contribution that biotechnology is making and might make in the future to the solution of these problems, to look first at the capability available before attempting to assess the performance and intentions of the industry.

4.1 CAPABILITY

Whilst the Green Revolution has not been effectively implemented, for example in Africa, and whilst further gains in productivity might be incrementally obtained through better farming methods, it does seem that further huge increases in potential maximum yields are unlikely to be obtained through standard methods.[33] A radically different approach to plant breeding has therefore been proposed. This involves using modern molecular maps to help unlock the genetic potential of wild species which have been assiduously collected into seed banks around the world.[34]

Tanksley and McCouch, two of the workers involved, argue that domestication of our main crops occurred quite rapidly under strong human selective pressures:

"...Certain traits, such as nonshattering of seeds, compact growth habit, or loss of germination inhibition, would have been selected by early agriculturalists. Selective propagation of lines containing these favourable mutations would have resulted in a progressive narrowing of the genetic base of subsequent populations."

Modern plant breeding over the last century has further narrowed the genetic base:

"...Because new varieties are usually derived from crosses among genetically related modern varieties, genetically more variable, but less productive, primitive ancestors are excluded..."

An example of the result is that virtually *all* modern soybean varieties used in the USA can be traced back to a dozen strains from a small part of China. Such inbreeding leaves crops very susceptible to pests and so, following the Southern corn leaf blight disaster in the United States in 1970, a worldwide effort was made to collect seeds of our 100 crop species and their wild relatives. There are now some 700 seed collections worldwide which hold some 2.5 million entries (Table 7).

TABLE 7. Estimated number of seed bank entries worldwide for selected crops *

Crop	Entries	Collections of 200+	Percentage of wild species
Wheat	410,000	37	60
Rice	215,000	29	10

Maize	100,000	34	15
Soybean	100,000	28	30
Potato	42,000	28	40
Tomato	32,000	28	70
Cotton	30,000	12	20

* From reference 34

Unfortunately, the use of this resource has been rather limited to date. This is the result of the prevailing plant breeding paradigm. Tanksley and McCouch believe that crossing the wild species with modern varieties will only show up simple dominant traits - for example, in resistance to disease - in the phenotype (manifest characteristics). Yet yield is clearly conditioned by many genes so they think that:

"...it is unlikely that modern cultivars have the best alleles for yield at all yield-related loci. Many beneficial alleles have undoubtedly been left behind because of the bottlenecks imposed by domestication coupled with years of modern breeding..."

They therefore argue for a different strategy based on searching for desirable genes through the use of linkage maps based on molecular markers.

There have clearly been huge advances in our understanding of plant genomes in recent years[35] and further progress is highly probable.[36] Tanksley and McCouch report that it has been possible to study:

"...the effects of the individual loci that control a quantitatively inherited trait (quantitative trait loci, or QTLs)..."

This has shown that different loci have very different effects and that a great deal of the genetic variation in a population is related to just a few QTLs. Furthermore, the phenotype is not a good predictor of the genetic potential. Thus in crosses of high and low yielding rice:

"...While the high-yielding line often does contain a great number of positive alleles at the loci associated with yield, there are almost always some loci for which the inferior parent contributes a superior allele."

They therefore argue that a search for such alleles using modern methods could produce dramatic improvements in yield.

Their results for the tomato show:

"...specific QTLs from the wild species *Lycopersicon hirsutum*...that outperform the original elite variety by 48, 22, and 33% for yield, soluble solids content, and fruit color, respectively..."

These results were confirmed under varying conditions in different parts of the world and similar results have been obtained for rice. Tanksley and McCouch look forward to the cloning of genes underlying yield-enhancing QTLs and then they suggest:

"...We can expect that such gene-based searches will ultimately be extended beyond the realm of sexually compatible species so that homologous genes from other genera and families can be tested for optimum performance by a transgenic approach."

This example demonstrates clearly that plant biotechnology[37] and the industry related to it[38] have considerable capacity for assisting in the task of feeding the world in the early decades of the next century.

The revolution in biotechnology has clearly had a major impact on all three aspects of health care: prophylaxis, diagnosis and therapy.[38] The underlying reason for this, of course, is the growing understanding we have of the mechanisms involved at a molecular level. This can be seen very clearly in the way that major research organisations[39] now set out their strategy for vaccine development (Table 8).

Probably the best single example of the impact of genetic engineering on vaccine development is the hepatitis B virus (HBV) subunit vaccine.[40] This was developed after vaccine manufacturers began to examine new routes of production when the advent of HIV forced a move away from the use of blood products:

"...The most appropriate system...proved to be expression of the surface antigen in yeast (*S. cereviseae*). The production of surface antigen lipoprotein particles allowed the development of the first licensed recombinant vaccine for human use....HBsAg particles were shown to be innocuous and have high protective efficiency..."

TABLE 8. Strategy for vaccine development *

"...First we need to understand better the protective immune responses to specific organisms..."

"...Secondly, we need to determine which are the most appropriate molecules to incorporate into a vaccine. Here the recent advances in molecular genetics have proved invaluable, providing the tools to help us home in on the right target molecules, and to analyse how variable an organism's structure can be."

"Thirdly, we need to understand the evasion strategies of micro-organisms, so that vaccines do not unwittingly incorporate molecules which interfere with the generation of effective immune responses..."

"...Finally, vaccines must generate long-lasting protection and, of course, be safe."

* From reference 39

The licence for general human use was granted in the US in 1986. It is important to note that modern approaches have also had a significant impact on the development of animal vaccines for severe problems of agricultural production such as foot and mouth disease, pseudorabies virus and Newcastle disease of poultry.[40]

A recent general review of trends in preventive medicine[41] demonstrated the wide range of approaches being taken - from subunit vaccines and naked DNA through to microsphere encapsulation - in the development of new vaccines (Table 9).

TABLE 9. Some vaccines in development *

Disease /Pathogens	Technology
Measles	Subunits in vaccinia; attenuated virus; naked DNA.
Cholera	Inactivated and attenuated pathogens; subunit combinations.
Shigella	Bacterial vector; subunits and conjugates.
Tuberculosis	Mycobacterial vector; subunit.
Malaria	Subunit and naked DNA.
Tetanus	Single dose: toxoid in microspheres.

* From reference 41

Clearly, as the strategic development plans of the various steering committees of the World Health Organisation's Programme for Vaccine Development show, there are enormous opportunities for the application of modern biotechnology to the improvement of health in the developing world. The crucial impact of new technology is clear from a recent detailed review, 'Vaccines and World Health', which noted:[42]

"From the late 1980s, research and development on vaccines has been revived in part because of the recognition that new biomedical technologies might improve existing vaccines and create novel products to protect against currently unvaccinable diseases....Much reliance is placed on technology to overcome universally recognised shortcomings in the current world vaccine picture..."

The question, of course, is whether such hopes and expectations can be realised.

Even for the specific issue of vaccines, however, what can be done is not just limited to research, properly directed, in the developed world. As the World Health Organisation makes clear,[43] almost half of the world's vaccine production takes place in developing countries and there is much that could be done to improve the quality of this production (Table 10).

TABLE 10. Vaccine supply and quality *

Almost 4 billion doses of vaccines are used in national immunization programmes each year.
Vaccines used in developing countries are either supplied through UNICEF, bought from another country or locally manufactured. Nearly half the world's vaccine supply is presently produced in developing countries for local use.
Vaccines supplied by UNICEF and many industrialized countries are generally of high quality.
Manufacturers in developing countries do not always produce vaccines under conditions which ensure safety, potency and efficacy of the products. Low quality vaccines means lives and programme reputations are put at risk.

* From reference 43

There are considerable difficulties with ensuring the quality of products through 'good manufacturing practice' (GMP):

"...GMP is a system of working which, once it has been established how to make a safe, potent, and effective product...guarantees that the product is always made the same way..."

There are also severe difficulties with the national control systems in many vaccine-producing countries.

TABLE 11. WHO requirements for adequate national control of vaccine manufacture *

- License vaccines according to published requirements.
- Evaluate clinical performance of the vaccine.
- Control and release each batch lot of vaccine individually.
- Perform laboratory testing.
- Monitor vaccine performance by post-marketing surveillance.
- Inspect manufacturing facilities and processes regularly.

* From reference 43

Many of the required functions are not adequately carried out:

"...As part of strengthening the capacity of NCAs (National Control Authorities) in each country which produces vaccines, WHO conducted an inventory of their activities. Of 34 countries visited to date, only 18 performed all six functions. Half lacked one or more of the critical national control functions deemed necessary to guarantee vaccine quality..."

It is noteworthy that it is precisely in regard to co-operation on such issues that Brazil has argued that Article X of the Biological and Toxin Weapons Convention might usefully be developed.[44]

4.2 INTENTIONS

Whilst it is not possible to discover the future intentions of an industry as vast and diverse as biotechnology today,[38] it should be possible to examine the recent history and present performance of specific parts of the industry in order to assess its likely future actions in these particular spheres of operation.

Despite the clear need for enhanced funding of agricultural biotechnology devoted to the needs of the developing world, it appears that the very reverse is happening. Overall, "Public agricultural research funding has been declining for years".[33] A recent international report stated that, in constant 1985 dollars:

"...global research spending doubled between 1971 and 1991, from $7.3 billion to $15 billion. But the average annual rate of increase fell from 4.4% in 1971-1981 to 2.8% in 1981-1991..."

Moreover, it is unlikely that private finance will fill the gap because for crops like wheat and rice, which self-pollinate, industry only gets to sell the seed once: "Economics thus drives private R&D to focus on drugs, pesticides, food processing and mechanisation".[33]

Unfortunately, the slowdown in funding has been particularly marked for international R&D. This is predominantly carried out through the Consultative Group on International Agricultural Research (CGIAR), a group of 43 private and public donors which supports 17 research centres around the world. Although CGIAR accounts for only two per cent of all agricultural R&D:[33]

"...its catalytic role is disproportionately important, especially in the Third World. The chief focus of many national research institutions in poor nations is developing local adaptations of CGIAR technologies..."

Yet CGIAR's budget has remained about level, at $315 million, since 1993. Furthermore, this money has been spread out over more institutions. The International Rice Research Institute (IRRI) based in the Philippines, for example, has lost almost a quarter of its $30 million budget in two years and has laid off 550 people - half its staff. The loss of R&D funding for agriculture is particularly marked in Africa where spending from all sources increased by 0.8 per cent in the 1980s, about one third of the worldwide average. Agricultural economists warn that whilst an integrated strategy *could* allow us to feed the expanding world population, failure to support the necessary R&D component could lead to failure of the entire project. That would appear to be the present prospect.

The need for greater funding of work on vaccines for the developing world has long been recognised. In a lengthy review for *Nature* in 1989 Barry Bloom wrote:[45]

"...Immunization is the most cost-effective weapon for disease prevention in developing countries, and new molecular and genetic technologies are making new types of vaccine feasible.... *What is lacking is the will to make the advances of modern biomedical science available to the poorest people of the world.*" (my emphasis).

Again, the problem is largely economic. An introduction to a 1994 major review of vaccines noted:[46]

"Companies can make money on a vaccine - if it's used in developed countries. Companies sell vaccines for one price in industrialized nations and provide them at a deep discount to groups such as UNICEF, which donate vaccines to poorer nations. *As a result, for diseases that don't afflict developed countries, there is scant commercial interest in vaccines...*" (my emphasis)

The author went on to say that malaria was a "telling example" which caused 300-500 million clinical cases and 2.7 million deaths each year.

Bloom's view had changed little by 1997 when he told *Science* that:[47]

"...The power of medical science to make a difference, were it to address the problems of people in the poorest countries, is quite remarkable, and yet there is no leadership that is linking biomedical science to global equity in health..."

Bloom was commenting on an Institute of Medicine report which argued that:[30]

"The United States is currently the global leader in biomedical science....The intellectual basis of its leadership derives principally from the longtime commitment of the US public sector to support fundamental research..."

This support was estimated to be $36-40 billion (0.6 per cent of GDP) in 1994. The report argued that:

"...The United States should...broaden the scope of its research and development activities to include health problems that impose the greatest burden of disease round the world, towards whose alleviation we can make important contributions. These problems include those infectious diseases that remain a major health burden in the developing world, particularly for children..."

There has, in short, been little change in funding over the last decade and it is hard to see a radical shift in priorities taking place today.

5. Conclusion

It is not, of course, the intention here to denigrate the work of those involved in co-operative international agricultural biotechnology,[48] or the tremendous achievements of the World Health Organisation, both in research[49] and in the field.[50] It is certainly not the intention either to deny the possibility of individual former offensive biological warfare facilities finding a role in the development and production of vaccines for, and in, the developing world.[7]

It is not possible to assess how a peaceful world economy might shape biotechnology across all the factors set out previously. However, given that the pharmaceutical industry will spend $25 billion next year on biological research and that a further $5-8 billion will be spent on biological research applied to agriculture, energy and environmental sectors of the economy,[51] it is reasonable to ask why so little is available to those in greatest need of the basic nutritional and medical resources.

The answer, it seems to me, lies in a fundamental misunderstanding, in the developed world, of our security interests. Despite the clear signs, for example in emerging and re-emerging diseases and forced migration, that problems in developing countries will affect us directly,[52] we all too often view their problems as involving choices over our charitable giving, not our vital interests. It is perhaps the question of our age as to how soon we shall wake up to the reality that Rosenberger warned of in his consideration of the world's food supply:[25]

"The real danger is to relegate the world food supply to the backwater of strategic studies. Strategists need to understand that the world food supply is a global challenge that bears most heavily on the peace and prosperity of the international system..."

And whilst world leaders could move this issue to the top of their agendas:

"...If they fail, their successors may have to deal with the problem 'when it comes to visit' as a major and enduring crisis in the early decades of the next century."

The rich world could, in short, provide the right context for the conversion of a war economy to a peaceful economy - within which facility conversions would be more likely to be successfully undertaken - if it changed its funding priorities to give more justice to the poor. The alternative is more conflict and a continuing development of national, science-based, high-technology war economies around the world.

References

1. Holum, J. D., Director, United States Arms Control and Disarmament Agency (1996) *Statement, Biological Weapons Convention Fourth Review Conference,* 26 November.
2. Office of Technology Assessment (1993) *Proliferation of Weapons of Mass Destruction: Assessing the Risks.* (OTA-ISC-559), United States Congress, Washington D.C.
3. Boulden, L. H. (1996) CIA, DIA provide new details on CW, BW programs in Iran and Russia, *Arms Control Today*, August, 32-33.
4. United Nations Security Council (1995) *Eighth Report of the Secretary-General on the Status and Implementation of the Plan for the Ongoing Monitoring and Verification of Iraq's Compliance with Relevant Parts of Section C of Security Council Resolution 687 (1991)*, S/1995/864, 11 October, United Nations, New York.
5. Leitenberg, M. (1996) *Biological Weapons Arms Control*, Project on Rethinking Arms Control, Center for International and Security Studies at Maryland School of Public Affairs, University of Maryland at College Park.
6. Harrington, A. (1995) Redirecting biological weapons expertise: Realities and opportunities in the former Soviet Union, *Chemical Weapons Convention Bulletin* **29**, 2-5.
7. Rimmington, A. (1996) From military to industrial complex ? The conversion of biological weapons' facilities in the Russian Federation, *Contemporary Security Policy* **17** (1), 80-112.
8. Leitenberg, M. (1994) The conversion of biological warfare research and development facilities to peaceful purposes, in E. Geissler and J. P. Woodhall (eds), *Control of Dual-Threat Agents: The Vaccines for Peace Programme*, SIPRI Chemical and Biological Warfare Studies, No. 15, Stockholm International Peace Research Institute, Stockholm.
9. Schofield, S., Dando, M. R. and Ridge, M. (1992) *Conversion of the British Defence Industries*, Peace Research Reports No. 30, Department of Peace Studies, University of Bradford, UK.
10. Andreev, Y. V. (1997) End of Russian Conversion? Paper presented at the 47th Pugwash Conference on Science and World Affairs, Lillehammer, Norway.
11. Borko, Y. (1997) Economic transformation in Russia and political partnership with Europe, in V. Baranovsky (ed.), *Russia and Europe: The Emerging Security Agenda.* Oxford University Press, Oxford (for SIPRI).
12. Dando, M. R. (1996) New developments in biotechnology and their impact on biological warfare, in O. Thränert (ed.), *Enhancing the Biological Weapons Convention.* Dietz Verlag, Bonn.
13. Toffler, A. and Toffler, H. (1993) *War and Anti-War*, Little, Brown and Company, New York.
14. Schofield, S. (1997) *Peace and Prosperity - Transition from Advanced Military Economy to Advanced Civil Economy*, Mimeo, Department of Peace Studies, University of Bradford, UK.
15. Southward, P. (1985) *The UK Defence Industry*, Peace Research Reports No. 8, University of Bradford, UK.
16. Schofield, S. (1986) *Employment and Security - Alternatives to Trident*, Peace Research Reports No. 10, University of Bradford, UK.
17. Rigby, A. (1988) *A Life in Peace: A Biography of Wilfred Wellock*, Prism Press, Dorset.
18. Woodhouse, T. (1986) To live our lives so as to take away the occasion for war: some observations on the peaceful economy, in G. Chester and A. Rigby (eds), *Articles of Peace: Celebrating Fifty Years of Peace News*, Prism Press, Dorset.
19. Wolffe, R. (1997) Quaker spirit on the wane, *Financial Times*, 20/21 September, p. 3.
20. Melman, S. (1985) *The Permanent War Economy* (revised edition), Simon and Schuster, New York.
21. Renner, M. G. (1988) Conversion to a peaceful economy: Criteria, objectives and constituencies, *Bulletin of Peace Proposals* **19** (1), 129-141.
22. Brecher, J. and Costello, T. (1994) *Global Village or Global Pillage: Economic Reconstruction from the Bottom Up*, South End Press, Boston, Mass.
23. Childers, E. (1994) Introduction, in E. Childers (ed.), *Challenges to the United Nations: Building a Safer World*, Catholic Institute for International Relations/St. Martins Press, New York.
24. Leader (1996) Feeding the world, *The Economist*, 16 November, p. 20.
25. Rosenberger, L. F. (1997) The strategic importance of the world food supply, *Parameters*, Spring, 84-105.
26. Maitland, A. (1996) Comment and analysis: The wheat from the chaff, *Financial Times*, 17 October.
27. Jamison, D. T. *et al.* (1993) *Disease Control Priorities in Developing Countries*, Oxford University Press, Oxford (for the World Bank).
28. Hull, H. F. and Aylward, R. B. (1997) Ending polio immunization, *Science* **277**, 8 August, p. 780.[
29. Russell, P. K. (1993) The Children's Vaccine Initiative: Technologic opportunities and management challenges, *Annals of the New York Academy of Sciences* **700**, 21 December, 1-7.

30. Board of International Health (1997) *America's Vital Interest in Global Health*, Institute of Medicine, National Academy Press, Washington D.C.
31. Lederberg, J. *et al.* (eds) (1992) *Emerging Infections: Microbial Threats to Health in the United States*, Institute of Medicine, National Academy Press, Washington D.C.
32. Morse, S. S. (ed.) (1993) *Emerging Viruses*, Oxford University Press, Oxford.
33. Mann, C. (1997) Reseeding the Green Revolution, *Science* **277**, 1038-1043.
34. Tanksley, S. D. and McCough, S. R. (1997) Seed banks and molecular maps: Unlocking genetic potential from the wild, *Science* **277**, 1063-1066.
35. Mazur, B. J. (1995) Plant biotechnology R&D - the next ten years, *TIBTECH* **13**, September, 313-323.
36. Cohen, J (1997) Corn genome pops out of the pack, *Science* **276**, 1960-1962.
37. Jones, M. G. K. and Lindsey, K. (1993) Plant biotechnology, in J. M. Walker and E. B. Gingold (eds), *Molecular Biology and Biotechnology*, Royal Society of Chemistry, London.
38. Moses, V. and Moses, S. (1995) *Exploiting Biotechnology*, Harwood Academic, Switzerland.
39. Beverley, P. (1996) A jab in time, *Medical Research Council News*, Winter, 16-18.
40. Mackett, M. (1993) Vaccination and gene manipulation, in J. M. Walker and E. B. Gingold (eds), *Molecular Biology and Biotechnology*, Royal Society of Chemistry, London.
41. Beardsley, T. (1995) Trends in preventive medicine, *Scientific American*, January, 68-72.
42. Bosch, P. (1994) *Vaccines and World Health: Science, Policy, and Practice*, Oxford University Press, Oxford.
43. Global Programme for Vaccines and Immunization (1997) *Highlights VSQ (Vaccine Supply and Quality) Documents*, World Health Organisation, Geneva, Website update, 17 October.
44. Brazil (1994) *Strengthening the BWC: Elements for a Possible Verification Regime*, Special Conference of the States Parties to the Convention on the Prohibition of the Development, Production and Stockpiling of Bacteriological (Biological) and Toxin Weapons and on their Destruction, BWC/SPCONF/Wp.4, 21 September, Geneva.
45. Bloom, B. R. (1989) Vaccines for the Third World, *Nature* **342**, 115-120.
46. Cohen, J. (1994) Bumps on the vaccine road, *Science* **265**, 1371-1373.
47. Cohen, J. (1997) Infectious diseases: Congress readies injection of funds, *Science*, **277**, 760.
48. Subrahmanyam, D. (1992) *Peaceful Uses of Biotechnology: Role of the ICGEB*, Ad Hoc Group of Governmental Experts to Identify and Examine Potential Verification Measures from a Scientific and Technical Standpoint, 3 April, United Nations, Geneva.
49. Bektimirov, T. *et al.* (1990) Vaccine development perspectives of the World Health Organisation, *J. Medical Virology* **31**, 62-64.
50. Hall, H. F. (1997) Pax Polio, *Science* **275**, 40-41.
51. Simpson, K. (1997) The impact of developments in biological science on biomedical markets, paper presented to the NATO Advanced Study Institute, 'New Scientific and Technological Aspects of Verification of the Biological and Toxin Weapons Convention (BWC)', 6-16 July, Budapest, Hungary.
52. Rogers, P. (1997) Reviewing Britain's security, *International Affairs* **73** (4), 655-669.

REDIRECTING BIOLOGICAL WARFARE CAPACITY TO INTERNATIONAL HEALTH BIOTECHNOLOGY

ROBERT B. CRAVEN, M.D.
Division of Vector-Borne Infectious Diseases National Center for Infectious Diseases Centers for Disease Control and Prevention Public Health Service U.S. Department of Health and Human Services
P.O. Box 2087 Fort Collins, CO. 80522, USA

The World Health Organization's (WHO) World Health Report 1997 presents a global picture of the unremitting importance of infectious diseases and particularly of their impact on developing nations. Worldwide, 17 of 52 million deaths (33%) in 1996 were due to infectious diseases. Forty of the 52 million deaths (77%) were in developing countries. Of the 40 million deaths in developing countries, 43% were due to infectious causes. Pneumonia killed 3.9 million people in 1996. Tuberculosis killed 3 million. Diarrheal diseases killed 2.5 million. HIV/AIDS killed about 1.5 million and 29.4 million children and adults were infected by HIV at the end of 1996.

To share with you the immediacy and scope of infectious disease problems around the world, I have compiled some data on outbreaks during 1996 and 1997 from data provided by WHO and by the Internet list server, ProMED. Between April and August of 1997, 17,592 cases of cholera with 910 deaths (5%) were reported from outbreaks in 12 African countries as well as in Iran, Pakistan and Ukraine. Meningococcal meningitis outbreaks in 18 African countries from January through August, 1997, produced reports of 55,885 cases. In six of these countries: Burkina Fasso; Gambia; Ghana; Mali; Niger; and Togo, 5,783 children died of meningococcal meningitis. A partial list of other infectious disease outbreaks which occurred in 1996 or 1997, but for which data are presently incomplete include anthrax, Congo-Crimean hemorrhagic fever, Coxsackie myocarditis, diphtheria, Japanese encephalitis, Lassa fever, leptospirosis, louse-borne typhus, measles, monkey pox, Murray Valley encephalitis, onyong nyong fever, plague, St. Louis encephalitis, typhoid fever, yellow fever (YF) and West Nile encephalitis.

For many of these diseases, e.g., diphtheria, Japanese encephalitis, measles, plague, typhoid fever and yellow fever, vaccines are available. Among this group of vaccines, the failure to use the live, attenuated yellow fever vaccine is worthy of particular scrutiny. From 1986 through 1995, over 23,000 cases of YF were reported to WHO. During this period, the African continent reported 21,638 cases and 5107 deaths (24%). In the same period, South American countries reported 2018 cases and 1391 deaths (69%). Under reporting of YF cases is notorious and incidence studies have

E. Geissler et al. (eds.), Conversion of Former BTW Facilities, 45-51.
© 1998 *U.S. Government. Printed in the Netherlands.*

consistently shown that only 1/ 80th or less of actual cases are reported. Why are so many people sick and dying from a disease for which a safe and effective vaccine exists?

For YF vaccine and for many other vaccines, the major impediment to use is cost. The Pasteur Institute YF vaccine is the most frequently used and least expensive YF vaccine in the world. Depending on how it is packaged and the quantity purchased, the price ranges from 0.14 to 0.17 USD per dose. This seems a minimal amount by industrialized world standards; but, in developing countries with limited hard currency to spend, this is 3 or 4 fold more expensive per dose than most childhood vaccines. In African countries in which I have worked to contain YF epidemics, Ministers of Health have all commented that if they were to include YF vaccine in their childhood immunization programs they would be forced either to drop some other vaccine such as measles or to reduce the number of children immunized. Since measles is a nearly constant severe threat to children, they opt instead to not use YF vaccine with the rationale that disease occurrence is episodic in most cases.

Often vaccine manufacturers are accused of corporate greed and overpricing. We must, however, be aware that vaccines can be very expensive to produce. First are the research and development costs. Even if the vaccine manufacturers do not incur those costs, then they must pay royalties to the patent holder. Next, a vaccine must undergo safety and efficacy testing prior to being licensed for production and sale. It is not unusual in industrialized countries for this process to consume 10 or more years. Manufacturers may also need to purchase insurance for their vaccine to protect themselves against litigation from recipients who feel they were harmed in some way by the vaccine. Actual production of the vaccine will entail additional costs; and, if the vaccine requires refrigeration, then storage and shipping will also add to the final cost.

The extremely high cost of bringing a vaccine to market means that companies will only develop vaccines which they can sell in high volumes over extended periods of time. Otherwise they have no hope of profit. Thus a decision to undertake commercial vaccine production means that there is a sufficiently large market and a price per dose can be charged which will both cover expenses and yield profit. An infectious disease which occurs episodically in countries unable to pay high prices per dose of vaccine is not going to attract the development interests of vaccine manufacturers. Thus, availability of vaccine depends not upon human needs but upon potential for profit.

Additional considerations for use of a vaccine are efficacy and logistics. A vaccine may have a reasonable price per dose, but multiple doses of it are required for primary immunity. Killed vaccines such as Japanese encephalitis and plague are examples of vaccines requiring multiple doses and thus causing a high final price per immunized recipient both in price of vaccine and costs of administering multiple doses. Does the vaccine produce acceptable immunity in the local recipients? Rotaviruses produce fatal cases of diarrhea in children of the developing world. For reasons presently obscure, candidate rotavirus vaccines so far produce encouraging immune responses in children of the industrialized world but are not equally immunogenic in children of the developing world who are most at risk of fatal infections.

Finally, vaccines and especially those for viral infections do not function optimally in situations where rapid onset of protective immunity is desirable. When YF epidemics occur, curtailing them involves mass immunization of the human population and attempts to reduce the populations of vector mosquitoes that transmit the disease. Immunization, however, requires a latent period of some 10 days to 2 weeks before people have actual immunity. Prior to this they are susceptible to being infected and becoming ill. Another example of how the delay in development of immunity presents problems is demonstrated by travelers. People traveling to areas where they need vaccines to protect them against endemic diseases frequently wait until right before departure to secure their needed vaccines. These individuals on a typical 1 to 2 week journey will be returning home before they develop an effective immune response.

1. Applications of Molecular Technology to Achieving Immunity against Episodic Infections

Is it possible then to mass produce biological products that will provide rapid or long lasting immunity at a low price per dose with a single administration that can be easily stored? Could these same products also function to protect against biological toxin warfare (BTW) agents and thus make them obsolete as offensive weapons? An affirmative answer appears to be available to both questions. The combination of two presently available molecular technologies: human monoclonal antibodies and DNA vaccines may well provide the solutions to the problems raised above of delivering immunity against infectious diseases in the developing world. And we may find in the process that scientific personnel and facilities formerly utilized in BTW research and production may be particularly appropriate to developing this technology.

Are there a sufficient number of candidate vaccine needs to support conversion of BTW facilities to commercially viable production sites for niche markets? Below is just a partial "shopping list" of vaccines (active and/or passive) which, if produced using modern, inexpensive technology, would be in international demand from focal and widely distributed areas. Such development of course presupposes international financial support for development of technology and conversion of facilities. Some good examples are:

- Malaria
- Yellow fever
- Dengue
- Japanese encephalitis
- Venezuelan equine encephalitis
- Arenaviruses (Lassa, Junin, Machupo, Sabia)
- Filoviruses (Ebola)
- Tuberculosis
- Plague
- Tularemia
- Cholera
- Many more

Human monoclonal antibodies can be produced in two ways: 1) from the lymphocytes of immune humans or 2) by "humanizing" antibodies produced from the lymphocytes of mice (murine antibodies). Both techniques have been successfully employed and both have advantages. Monoclonal antibodies derived from humans are ready to be administered to provide virtually instant passive immunity as soon as they have been successfully produced. Also, in deriving monoclones from human donors, multiple monoclones effective against a variety of agents can be obtained from a single donor. For example, in one donor who has been immunized against a number of tropical infectious disease such as I have, it is theoretically possible to obtain from me alone lymphocytes producing antibodies against YF; Rift Valley fever; plague; eastern, western and Venezuelan equine encephalitis; and hepatitis A and B.

Murine monoclonal antibodies as a starting point have the advantage that they already exist against almost every known infectious agent. The problem they present is that, when administered to a human, they are quickly recognized as a foreign protein and elicit an immune response themselves which, of course, inactivates them. Thus, they have a very short effective life. In many cases, however, it is possible to genetically alter these murine monoclones to make amino acid substitutions in their antibodies that make them appear more human in structure.

The techniques for producing murine monoclonal antibodies are well known and need not be discussed here. The procedure for producing human monoclonal antibodies is newer and has largely been applied to cancer therapy until recently. The procedure begins by selecting a human donor known to be immune to the agent against which monoclonal antibodies are desired. Lymphocytes from this donor are the object and can be obtained from peripheral blood or most efficiently from a specimen of bone marrow.

Lymphocytes from the donor are removed from other blood or marrow components. The RNA they contain which codes for antibodies is then isolated, amplified and removed. This RNA for antibodies is then converted to complementary DNA for antibodies. This complementary DNA is then inserted into the genes of a bacteriophage. The bacteriophage is then allowed to replicate thereby producing a mixture of antibodies on its surface. To obtain the specific desired phage- antibodies from phage-antibodies directed at other infectious agents, the desired target virus is mixed with the phage- antibodies to form bound antigen-antibody complexes. The specific antibody producing genes can then be isolated, removed and purified. The purified genes are then introduced into an expression vector such as the bacteria, E. coli. Growth of such vector E. coli in an inexpensive fermentation process will yield stable, protective human antibodies that can be stored lyophilized for later reconstitution and injection to provide virtually instant immunity.

What are the potential uses for human or humanized murine antibodies? There are several important potential uses. First, they could be used as BTW deterrents. Stockpiles of monoclonal antibodies effective against BTW agents distributed around the world would serve as strong deterrents to employment of BTW agents effectively neutralizing the potential for their deployment. Secondly, human monoclonal antibodies (HuMAbs) would allow rapid intervention in natural epidemic situations. Populations at risk on the advancing edge of an epidemic could be given virtually instant immunity. Third, travelers entering endemic disease areas could be protected immediately prior to departure or even at the endemic risk destination. Fourth, HuMAbs may have a role in therapeutics as a means to reduce or eliminate infecting agents, particulary viral infections for which so few effective chemotherapeutic interventions are available. Fifth, HuMAbs, when combined with appropriate indicator systems, could be useful diagnostically for rapid antigen detection.

An additional area in which molecular techniques show promise for international health are DNA vaccines. Briefly, these are produced as follows. For the target infectious agent, the area where antibody attachment will neutralize the agent is identified. Then in the target infectious agent, the genes which code for this site that is vulnerable to neutralizing antibody attachment are identified. These identified genes are removed from the infectious agent and spliced into a bacterial plasmid. The bacterial plasmid is then inserted into a vector such as E. coli. The E. coli and its plasmid are then replicated in a fermentation process. The DNA plasmid is then separated from the E. coli and purified. The purified plasmid when injected into a vaccinee will then reproduce the target binding site from the original target infectious agent and induce immunity against that agent.

1.1 ADVANTAGES OF HUMABS AND DNA VACCINES AS BTW DETERRENTS AND PROMOTERS OF INTERNATIONAL HEALTH BY PRODUCTION IN FORMER BTW FACILITIES

1) The technology to produce HuMAbs and DNA vaccines is available, tested and shown to be practical.

2) Development and approval phases for HuMAbs and DNA vaccines should be shorter than for conventional vaccines.

3) Both approaches should provide protection not only from natural infectious diseases epidemics, but also should protect against most, if not all, BTW agents.

4) Production costs are low for both technologies.

5) Both types would have a long shelf life and would be portable without refrigeration under field conditions.

6) International support of development costs should allow final pricing to reflect more production costs than development costs.

7) Unlike inactivated vaccines, early evaluation of DNA vaccines suggest that they produce strong, long-lasting immunity.

8) HuMAbs produce virtually instant immunity useful in management of epidemics, indispensable in protecting against threatened use of BTW agents, and in providing protection for the growing numbers of travelers visiting areas with exotic diseases.

1.2 CONCLUSIONS ON REDIRECTING BTW FACILITIES TO PEACEFUL INTERNATIONAL HEALTH NEEDS

1) Molecular technology available now can produce active and passive vaccines, therapeutic and diagnostic agents.

2) Molecular technology applies equally well to deterring BTW use and to controlling the risk of exposure to natural infections.

3) Former BTW facilities and scientists, cooperating internationally, may be uniquely qualified to develop inexpensive solutions to the health problems of developing nations.

4) Once converted to peaceful health uses, former BTW facilities could become biotechnology training centers for the developing world and could become major centers of biotechnology transfer to those nations.

5) Improving health in the developing world is a key step to increasing economic progress in those countries.

6) Improved control of infectious diseases in the developing world is an investment in the health of the developed world.

7) Redirection of former BTW facilities to world health problems is an investment in international peace.

2. Discussion

Major discussion points raised were: 1) the strong likelihood that subsidies from international and national donors (e.g. UN, WHO, EEU, USA, etc.) rather than participation from the pharmaceutical industry would be the key to transforming existing facilities into centers for developing world health technology. The political process involved in obtaining these commitments would likely be longer than direct projects having profit potential with the existing international pharmaceutical industries. 2) World standards for facilities producing pharmaceuticals and biologicals were being promulgated by the World Health Organization. Meeting these facilities standards would be expensive for most existing BTW facilities. 3) Conventional disarmament wisdom has discouraged development of defensive measures in the face of potential success in disarmament, since development of successful defenses can trigger new rounds of developing effective offensive capabilities. It was pointed out, however, that in the course of developing active or passive vaccines for natural diseases, it was inevitable that an effective prophylactic against a BTW deployment was an additional and largely unavoidable end result. While signatory States to a BTW disarmament document might not be expected to attack each other with these weapons, terrorism concerns might make the availability of defensive measures a important additional assurance measure.

CONVERSION OF BTW FACILITIES: LESSONS FROM GERMAN HISTORY

ERHARD GEISSLER,
Max-Celbrück-Centrum für Molekulare Medizin
13122 Berlin, Germany

1. Introduction

The objective of the Biological and Toxin Weapons Convention (BTWC) is the total prohibition of the development, production, storage, stockpiling, acquisition, and transfer of biological and toxin weapons. The prohibition in Article I is broad and States Parties are required by Article II to destroy or divert to peaceful purposes all agents, toxins, weapons, equipment and means of delivery designed to use biological agents or toxins for hostile purposes or in armed conflict. As the dual purpose nature of the facilities which develop, produce, stockpile or weaponize biological agents and toxins has long been recognized, the BTWC unlike the Chemical Weapons Convention (CWC) does not specifically address what should happen to past BTW facilities. In contrast, facilities for development and production of chemical weapons (CW) were generally built to produce CW agents and the CWC consequently addresses the destruction of such facilities as well as their temporary conversion to CW destruction facilities.

Nevertheless, even though the BTWC does not specifically address past BTW facilities, it is clear that such facilities should cease all activities associated with the past offensive BTW programme. There are clearly several options: the facility can at one extreme be destroyed completely so that the buildings are razed to the ground through to being converted into a facility for permitted purposes whether for BTW defense or for civilian applications.

In this chapter, the lessons that can be learnt from German history about its facilities which had interests related to biological warfare or biological defence are addressed.

2. Historical perspectives

International legislation in the field of biological and toxin warfare starts with the adoption of the Geneva Protocol in 1925 [1]. Since the Protocol deals exclusively with the *use* of bacteriological (biological) weapons in war and not with research, development and production, it is obvious that aspects of termination of BTW agents development and production are not addressed by that treaty.

E. Geissler et al. (eds.), Conversion of Former BTW Facilities, 53-66.
© 1998 *Kluwer Academic Publishers. Printed in the Netherlands.*

The situation changed in 1972 with the adoption of the BTWC which prohibits the development, production and stockpiling of bacteriological (biological) and toxin weapons [2]. Nevertheless conversion is not specifically mentioned in the BTWC. The only one article of the BTWC, which is implicitly related to conversion, is Article II. Article II states that "each State Party to this Convention undertakes to destroy, or to *divert* to peaceful purposes [...] all agents, toxins, weapons, equipment and means of delivery specified in article I of the Convention [i.e. "microbiological or other biological agents, or toxins {...} of types and in quantities that have no justification for prophylactic, protective or other peaceful purposes" and "[weapons, equipment or means of delivery designed to use such agents or toxins for hostile purposes or in armed conflict"], which are in its possession or under its jurisdiction or control [...]". Facilities where such agents are developed, produced, stockpiled and/or weaponized are not explicitly mentioned in that article or elsewhere in the Convention [emphasis given].

Agreement of the BTWC was, *inter alia*, the result of joint activities of governmental and non-governmental bodies and of numerous scientists. Termination of the work of BTW facilities was explicitly mentioned neither in the BTW nor in the majority of publications dealing with biological arms control, much less was the term "conversion" used. Swedish microbiologist Carl-Göran Hedén was among those few who considered the redirection of military microbiology. In a review published in 1967 he mentioned that military microbiological establishments would, "unlike other military establishments, [...] require no major reorganization in order *to switch* most of their activities to a highly constructive path. [...] The potential of military microbiology is illustrated by its civil 'spin-off': vaccines against Newcastle disease, fowl plague, and rinderpest, improved diagnostic and therapeutic methods, and so on" [3] (emphasis added).

The first obvious use of the term *conversion* was used by the participants of the 10th International Congress of Microbiology in 1970. In the closing session the microbiologists passed a resolution urging that, *inter alia*, "all installations where established microbiological programmes have been carried out expressly for offensive or defensive biological warfare purposes be *converted* to peaceful uses, if possible with international participation, and that ... all stockpiles of biological weapons should be destroyed as soon as possible". At the same time the participants expressed their "believe that existing installations for military microbiology could usefully be *converted* to any of the following uses: Applied environmental microbiology. Biogeochemical transformations. Conservation of soils. Crop productivity and biological nitrogen fixation. Production of proteins and other food substances. Production of enzymes and hormones. Purification and recycling of sewage for drinking water. Rapid diagnosis of infectious diseases. Improved means of mass-vaccination" [4] (emphasis added).

One year later Hedén referred to that resolution at an international conference on NBC disarmament and supplemented the list of potential fields for use of former BW facilities by adding "Problems of viral carcinogenesis. Problems of molecular biology. Microbiological pest control. etc" [5]. However Hedén did not use the term "conversion" but discussed the "*redeployment* of military microbiology towards civilian needs" (emphasis added). In a SIPRI monograph published ten years after SIPRI's six-volume CBW series to evaluate the present state of BTW matters in the mid-1980s [6] conversion or other means of termination of BTW facilities was not mentioned.

When it turned out that the BTWC needed strengthening the participants of the Second Review Conference decided in 1986 to strengthen the Convention by adoption of a set of Confidence-Building Measures (CBMs). Conversion of former BTW facilities, again, was not explicitly addressed and no information on conversion was requested. The same held true for the extension and amendment of the CBMs at the Third Review Conference in 1991. In SIPRI's analysis of the results of the first three rounds of CBMs conversion or others means of termination of BTW facilities again was not mentioned (but see below) [7] .

However, at least one codepositary country, the United States, voluntarily reported at the Second Review Conference that "facilities previously used for development, production or stockpiling of biological and toxin weapons were converted to peaceful purposes well in advance of the time specified by the convention" [8]. It is interesting that the US chose to use the term "converted" in its statement at it demonstrated the understanding of at least one co-depositary that it was implicit that States Parties to the BTWC are requireed to convert past BTW facilities.

It took another six years that conversion of former BTW facilities was discussed at length. In 1992 an international programme of development and use of vaccines against dual-threat agents (DTAs) was proposed to, *inter alia*, provide transparency in the R&D and production of vaccines, including vaccines for military purposes, to help to prevent the development and use of biological and toxin weapons for offensive purposes, to contribute to health protection, to promote peaceful international co-operation and, last but not least, *to contribute to the conversion of BW and TW facilities*. A number of the topics debated at the NATO ARW on conversion and discussed in this volume were also addressed when the Vaccines for Peace programme considered the possible involvement of former BTW facilities in that programme [9]. The subjects addressed included: how could the transfer of BTW technology and/or equipment for prohibited purposes be prevented; how could the conversion of the approximately 70 institutes of the former USSR involved in the Soviet "Biological Defense Research Program" to facilities engaged solely in peaceful programmes be facilitated; how can future employment be provided for the scientists and technicians of these facilities so that their skills are only used for national and international health care purposes?

Since concern was expressed about the possible danger of re-use of converted former BTW facilities for prohibited purposes it was emphasized that "assurance would have to be given that the converted facilities could not be reconverted [1] to offensive biological warfare facilities and that their personnel would not become involved in

[1] It was emphasized by Milton Leitenberg and other participants of the Budapest workshop that "reconversion" is a misleading term. A possible process of transforming a converted former BTW facility into an active BTW facility should be described instead as "reversion".

proliferation." [10] The Vaccines for Peace proposal was evaluated by experts in a series of private international meetings. These culminated in a confernce in Biesenthal near Berlin, Germany, in September 1992. At this conference Milton Leitenberg dealt at length with the problems involved in conversion of former BTW facilities [11]. Describing the conversion of US and British BW facilities Leitenberg quoted a report submitted as early as 1972 to the U.S. Congress on "problems associated with *converting* [BTW] defense research facilities" [12].

Recently, conversion of a BioMedPreparat facility of the former Soviet Union located in Stepnogorsk, Kazakstan (see Rimmington, this volume) was described as a "BW dismantlement" project and as "demilitarization" of the facility [14].

3. The termination of German BTW facilities in 1945

There are several reasons why BTW facilities may cease operation:

i) Arms reduction and disarmament as consequence of unilateral or multilateral decisions and agreements. Examples are the 1969 and 1970 decisions of the US Administration to stop offensive BTW activities and to adopt the BTWC.

ii) Military defeats or partial defeats of countries previously running BTW facilities independent of the existence of the BTWC or of whether the said state was a State Party to the BTWC. Examples are the defeat of France in 1940, of Germany and Japan in 1945, and of Iraq in 1991.

iii) In addition the activities of BTW facilities might be terminated by more or less complete destruction as consequence of enemy attack independent of the final outcome of the said war.

In this section the fate of German BTW facilities is discussed to determine what lessons can be learnt from German history about the termination of the work of German BTW facilities [for details see 14]. The fate of British, US, and Soviet facilities is described by other contributors to this book. In addition Thomas Stock deals in this volume with the conversion of former CW production facilities under the Chemical Weapons Convention.

3.1. WWI FACILITIES

It is not possible to consider the experience from Germany's World War I (WWI) defeat, although it is well established that BW agents were used by the Germans for hostile purposes already during 1915-17. German agents disseminated or attempted to disseminate *Bacillus anthracis* and *Pseudomonas mallei* among cattle and military horses and mules during WWI in Argentina, Romania, Spain, the USA and possibly also Scandinavia [15, 16]. At least the BW agents used for sabotage activities in Romania and Spain were delivered by the Ordnancy Department of the Section Politics of the General Staff ("Materialien-Verwaltung der Politischen Abteilung des Generalstabs"). It is not known with certainty, which facility provided the Ordnancy Department with the bacteria but there is reason to assume that the germs were supplied by the Bacteriological Laboratory of the Military Veterinary Academy, Berlin, Hannoversche Str. 27.

It is questionable whether that facility can be described as a BTW facility *sensu strictu* since its primary function was to take care of the health condition of military horses and other animals. If the laboratory was the supplier of the anthrax and glanders bacteria used for sabotage actions that activity was at most a byproduct of its main activities. In any case the activities of the laboratory were not terminated in 1918, not least because the German BW sabotage activities are not mentioned in the Versailles treaty (presumably because the BW sabotage activities remained marginal and were of no measurable success) in contrast to the German use of chemical weapons.

3.2. MILITARY BACTERIOLOGICAL INSTITUTE

In 1937, the secretary of the British BW committee circulated a document claiming that "on instructions from the Heeresverwaltung the M.B.I. (Military Bacteriological Institute) Berlin, under the direction of Dr. Hugo Stoltzenberg, is cultivating 'spore' bacteriae (anthrax bacteriae) in large quantities and storing them in dry condition" [17]. Numerous details were given on how the spores should be disseminated. Although it was, according to a document distributed a little later, a "doubtful report", issued only "under strong reserve" [18] the British were convinced in August, 1939, that they had reliable intelligence reports of German tests with anthrax-filled shells [19].

But that information was wrong (as discussed at length elsewhere [1][2]). After WWI and until the discovery of a French BW facility in 1940 there was *no* BTW facility run in Germany, not even a dual-use facility similar to the status the Bacteriological Laboratory of the Military Veterinary Academy obviously had in 1915-17 (see above)[3].

Only after the discovery of the French BTW programme BTW activities and especially biological defense preparations were initiated in Germany. Although several facilities were involved in such activities they were performed on a quite limited scale, especially Hitler had forbidden offensive BTW preparations and did not change his mind throughout the war.

It is also remarkable, that the leading bacteriological institutes, including the Robert-Koch-Institute, the "Reichsanstalt zur Bekämpfung der übertragbaren Krankheiten" (RKI) and the Bernhard-Nocht-Institut für Schiffs- und Tropenkrankheiten Hamburg, were not at all involved in such activities (but see

---

[2] However, some BTW agents such as *Bacillus anthracis* were studied in non-military institutes and pharmaceutical enterprises for peaceful health care purposes since BTW agents are dual-threat agents.

[3] If measures considered presently for strengthening the BTWC such as declarations and inspections would have been possible already in the late 1930s it could have easily been determined that Germany was not preparing biological warfare. Consequently it would have been not necessary for other states to prepare for BW defense or even retaliation measures.

below). The RKI was involved, however, in the production of vaccines against "war diseases" such as typhus and yellow fever.

Even more astonishing is that the Institute for Hygiene of the Waffen-SS headed by Joachim Mrugowsky was not involved BTW activities. This institute had a staff of about 200 persons at the end of the war, including 25 scientists, and one of its departments was a department of bacteriology and serology. Other departments were involved in the development of vaccines against typhoid and typhus as well as in research on arthropods as vectors for pathogens. Besides, Mrugowsky was directly responsible for testing of typhus and yellow fever vaccines produced by the Behringwerke Marburg and the RKI in experiments involving prisoners in concentration camps [20].

3.3. KLIEWE LABORATORY

Only after the Germans had discovered a French BTW facility, the "Laboratoire de Prophylaxie" in der Poudrerie Nationale du Bouchet in Vert-le-Petit near Paris, a bacteriological department of the Wehrmacht was established in autumn 1940. The facility, often referred to as "Kliewe laboratory", was not an autonomous institute but a small department of the Institute for General and Military Hygiene (*syn.* Institute for Hygiene and Bacteriology) of the Military Medical Academy (MMA) in Berlin. The space of the facility was quite limited. Its head, Professor Heinrich Kliewe had an office and - according to different reports - two to five rooms at his disposal, but was able to use laboratories of other departments of the Academy for certain experiments. His staff only included three to four scientists.

The work carried out by Kliewe was stimulated by his findings at the Laboratoire de Prophylaxie and by conclusions he drew from an evaluation on the intelligence reports on foreign BTW activities. A major part of his early experimental work was devoted to the evaluation and continuation of the French experiments on the combined activity of BW and CW agents. But, because of the limitation in lab space and the small size of the group, Kliewe's work was performed on a very small scale. The bacteriological experiments were carried out, for example, with an incredible low number of petri dishes certainly unsufficient to obtain reproducible results. In one of his reports he complained that "we do not possess the equipment required to work with pathogenic organisms on a large scale. It is recommended that the suggested studies be carried out immediately by a suitable bacteriologic-hygienic institute" [21]

When Kliewe was ordered in 1942 to confine himself entirely to the study of defensive measures, he initiated experiments with novel disinfectants including their application as aerosol. At least some of the results obtained in experiments using *Serratia marcescens* as test bacteria were published in the open literature, where ist was even mentioned that the experiments were carried out in the Institute for General and Military Hygiene of the MMA.

Since at least some of the activities performed in the Kliewe laboratory had clearly offensive intentions this facility can be described as one of the few German BTW facilities. Its fate after the war is not completely clear. In 1944 the laboratory was translocated to an evacuation site in Gießen, the Hessian Diagnostic Laboratory for Infectious Diseases (Hessisches Untersuchungsamt für Infektionskrankheiten) where Kliewe had been working already before WWII. Both places, the laboratories in Berlin and in Gießen, were severely damaged during the war. The Kliewe laboratory in Berlin was conquered by the Red Army together with the other buildings of the Military Medical Academy (22). Whether they found functioning equipment in the Kliewe facility is totally unknown. If there was any it presumably was dismantled similar to the equipment in other facilities located in Eastern Germany.

3.4. INSTITUTE FOR MICROBIOLOGY SACHSENBURG

The facility was founded in October 1943 in the Sachsenburg in the vicinity of Frankenberg near Chemnitz because it was expected "that Russian planes will disseminate plague agents over the German Reich. It was necessary, therefore, to take defense measures immediately and to establish a plague institute" [23]. Stabsarzt Dr. Hermann Gildemeister was appointed head of the institute. The main task of the institute was the production of plague vaccine and the conduction of animal experiments for plague diagnosis. In addition, experiments on tularaemia were performed in the Sachsenburg facility.

In October 1944 the staff consisted of about 20 persons. Despite of a lot of technical difficulties and despite a serious lack in manpower 60 000 charges and 76 litres of plague vaccine were produced in the first three months of 1945.

The testimony of Generalarzt Professor Schreiber in a letter to the Soviet government [24] and at the Nuremberg trial according to which the Sachsenburg facility "should be involved in bacteriological warfare preparations and should in addition to its actual task to produce vaccines for defense purposes also produce plague bacteria" was unequivocally wrong. The institute was clearly a pure BTW defense facility which might by run even today because no activities prohibited by the BTW Convention were performed there.

In early May 1945 the Red Army occupied the Sachsenburg and the institute was completely dismantled. Some of its rooms are used now as a museum[4].

3.5. PROVING GROUNDS

When the cooperation of the German Reichswehr with the USSR in development and testing chemical weapons--which did not include a cooperation in BW activities--was terminated in 1933, the Army Proving Ground ("Heeresversuchsstelle") Raubkammer/Munster-Lager was established in 1935 in an area of the Lüneburg Heath called "Raubkammer", North East of Munster [25]. Assigned to Wa Prüf 9, the CW Defence Section of the Army Ordnancy Office, the facility had to develop, test and

---

[4] The microbiological facility is so far not mentioned in the exhibition, not least because its existence was nearly totally unknown to the people living there. With the support of the present author the museum in future will include also a presentation dealing with the institute of microbiology and related topics.

weaponize CW agents. Part of the Raubkammer was used already in WW I for CW activities.

Attached to the Heeresversuchsstelle was a testing field for the Air Force. Experiments were performed in that field in cooperation with Wa Prüf 9. This testing ground was used by Kliewe's group for a few field trials conducted with *Serratia marcescens* and other simulants in early July 1942.

The Proving Ground is now used by the Federal Armed Forces (FAF). Besides being used as training area it harbors the FAF Defence Scientifc Agency for NBC Protection. Its Biology Section is engaged in B-defence (approximately 40 percent) and in environmental protection (approximately 60 percent). [26] The test field, hence, was converted.

An island in Lake Peipus in the occupied Soviet Union was used for a few experiments carried out by Drs. Hanns-Christoph Nagel, the BW expert of the Veterinary Inspectorate, and Kurt Stantien of Wa Prüf 9 with foot-and-mouth disease virus in 1942 and/or 1943.

Occasionally a "Testing ground East" ("Versuchsfeld Ost") or "bacteria field East" ("B-Feld Ost") is mentioned in German documents dealing with BW activities. It was planned to perform there experiments with simulants, foot-and-mouth disease virus, arthropods and weeds. The location of B-Feld Ost is unknown. It does not seem to be identical with the island in Lake Peipus mentioned above.

3.6. CENTRAL INSTITUTE FOR CANCER RESEARCH

The Central Institute for Cancer Research ("Zentralinstitut für Krebsforschung, Reichsinstitut an der Reichsuniversität Posen") was founded in 1942 as a dual-use facility. The institute was designed to become both a cancer research center as well as the major German BW facility. Its director Professor Kurt Blome, a dermatologist and cancer epidemiologist, was appointed the State Research Council's coordinator for cancer resarch and civilian BW activities[5]. The institute was provided with the monastery Nesselstedt about 7 km far from the centre of Posen as the main building. In addition space in the medical institutes of Posen University was used by some departments of the institute.

Departments for Physiology and Biology, for Bacteriology and Vaccine, for Obstetrics and Surgery, for Radiology, for Cancer Statistics and for Pharmacology were planned. In addition an animal farm in Bielen near Birnbaum should provide for experimental animals and house a tumour farm. In the Nesselstedt area five large buildings were planned. But there was very slow progress in the construction of the facility and none of the BW labs was completed when the Red Army conquered Posen in February 1945.

In contrast to other claims [27] much less than 86 persons belonged to Blome's staff. In addition to Blome only a few of the Nesselstedt scientists, including

---

[5] Military BW activities were coordinated by the Blitzableiter committee headed by colonel Walter Hirsch, the head of Wa Prüf 9.

bacteriologist Dr. Karl Josef Gross and, possibly, pharmacologist/toxicologist Dr. Hans Seel, were engaged in BW activities. Gross, formerly a staff member of Mrugowsky's Hygiene Institute of the Waffen-SS, was involved in the estbalishment of the BW sectors facility and in the preparation of experiments on *Yersinia pestis*. Seel, was involved in dual-use experiments on chemical pest control from aircraft.

According to Blome no BW activities work were performed there because not a single laboratory was ready for use. That statement was confirmed by the Polish public prosecutor H. Ciesluk. Ciesluk determined in 1951 that "none of the ... barracks had been completed, so that contrary to the statement made by Walter Schreiber [at the Nuremberg trail], no experiments could have been started yet in this institute" (emphasis given in Polish original) [28].

3.7. STATE RESEARCH INSTITUTE INSEL RIEMS

By far the largest dual-use facility utilized to a small extent for BW activities was the State Research Institute Insel Riems (Reichsforschungsanstalt Insel Riems]. The institute was founded in 1909-10 mainly to study and to treat foot-and-mouth disease (FMD). To prevent spread of the disease according to the obligations of German law the facility was established on the small Island of Riems located in the Baltic north of Greifswald. By the end of WWII, the institute included 1 000 square metres (s.qm.) laboratories, 5 500 s.qm. research staples, 1 900 s.qm. butchery rooms, 450 s.qm. workshops and a living quarter with 18 houses.

Professor Otto Waldmann; was director and president of the facility from 1919 to April 1948. During the war the total number of employees figured between 99 and 140 persons, among them about twenty scientists. The main task of the facility was the provision of immune sera and vaccines to counter FMD and other infectious animal diseases.

During the first years of WWII and the preceding period there are no indications for an involvement of the institute in BW activities with the exception that Hanns-Christoph Nagel, a former Riems researcher, was the BW expert of the Veterinary Inspectorate of the army (V In) for BW matters and served as V In's representative in the BW (Blitzableiter) committee.

But the facility was - although indirectly -- involved in offensive BW preparations, since the FMD virus used in open air experiment(s) by Nagel and Stantien (see above) had been produced on Riems. Hence the Riems institute can be regarded as a BTW facility.

In addition the institute was involved in defensive measures by provision of FMD vaccine and antiserum as well as by participation in the evaluation of outbreaks of FMD with respect to their possible origin by enemy attack. In the final stage of WWII the institute was requested to develop a rinderpest vaccine because of German concern that her enemies might use rinderpest virus for BW attacks. There are contradictory reports whether at least small-scale experiments with Rinderpest virus had been performed at Riems by Waldmann's deputy Erich Traub and his coworkers.

Between July and October 1945, the Riems facility was totally dismantled. Nevertheless the Soviet Military Administration in Germany requested as early as 1 January 1946 the reconstruction of the facility to provide vaccines and immune sera for FMD and other animal diseases as soon as possible. That means, the Riems facility was converted.

3.8. POTATO BEETLE RESEARCH STATION

In a directive on "The defense against BW" the High Command in 1942 ordered cooperation of Wa Prüf 9 with the Biological State Research Institute for Agriculture and Forestry, Berlin-Dahlem and the Ministry of Food and Agriculture "for protection of vegetation and plant food supplies against BW agents". One of the leading scientists of the State Research Institute, Professor Albrecht Hase, who has had in the past close relations to Nobel prinze winner Fritz Haber, Germany's leading WWI expert in chemical warfare, was specilized in pest control and an advicer to the army in CW matters at least in the 1920s. It is not known, however, whether the State Research Institute became directly involved in BW activities. Many documents of the facility were destroyed after the war. Those documents which escaped the destruction are now filed at the Federal Archives; they do not contain any reference to BW activities.

However, some BW activities were performed by a branch of the facility, the Potato Beetle Research Station in Kruft near Koblenz, which was established in May 1940 to coordinate the control of the natural spread of potato beetles. The director of the station was Oberregierungsrat Dr. Martin Schwartz. The station was a small facility consisting of a wooden four-room laboratory, two transportable shacks and two greenhouses. 18 persons were employed, among them 2-4 scientists. Although countering potato beetles was a serious civilian agricultural problem at least some of the studies at Kruft were coordinated not by the representative of the State Research Council responsible for agricultural research but by the coordinator of BW activities, Kurt Blome, and were registered under "Cancer Research" as a cover. This demonstrates that the facility was directly involved in BW activities.

It is unknown, however whether the station was directly involved in the preparation and execution of some open-air experiments carried out with potato beetles with offensive intentions or whether its tasks were restricted to defense measures only. Hence it is not clear whether the Potato Beetles Research Station can be described as a BTW facility *sensu stricto*.

By the end of 1944 the station was translocated to Mühlhausen in Thuringia because the Anglo-American troops were approaching the area of Kruft, and the number of employees war reduced to four. After the war Schwarz and his crew were requested by the Soviet Military Administration to re-establish a (civilien) potato beetle defense service in East-Germany. If his facility would meet the criteria of a BTW facility it was converted after the war.

3.9. INSTITUTE OF ENTOMOLOGY OF THE WAFFEN-SS AND POLICE

Another facility coordinated by Blome was the Entomologisches Institut der Waffen-SS und Polizei. Established in 1942 by order of Heinrich Himmler the institute was built in the immediate vicinity of the concentration camp Dachau near Munich. Dr. Eduard May, was appointed as its director. The institute was established as a facility of the SS Association "Ahnenerbe" [29].

The institute of entomology was a typical dual-use institute and may have even been planned originally as a facility to develop methods only for the control of fleas and other bugs and also for countering rats because Himmler was highly interested in these fields of hygiene. The relatively low priority of the institute becomes obvious not least by the fact that it took about two years to establish it and to provide it with equipment. Even after the institute became involved in BW activities it continued to suffer from a lack of support.

In September 1943, May met Blome to discuss measures of defense against potato beetles and other anti-plant insects in case the enemy would use such weapons as countermeasures *after a similar German attack*. In addition, May carried out experiments regarding the artificial spreading of mosquitos infected with malaria.

There were also connections between May and Blome's co-worker Gross presumably regarding work on *Yersinia pestis*.

As late as in March 1945, only days before the allies reached Dachau, May and his superior considered an enlargement of the Institute of Entomology and requested additional lab space for their experiments to be carried out "to evaluate the possibilities of biological warfare".

The facility clearly did not survive WWII[6] but details of its fate are unknown.

3.10. KAISER WILHELM INSTITUTE FOR HORTICULTURE RESEARCH

In the field of anti-plant agents, the use of weeds to interfere with crop production was considered by those few German activities who had been in favour of BW. Experiments to evaluate the possibility to use the seeds of weeds as BW agents were carried out at the Kaiser-Wilhelm-Institut (KWI] für Kulturpflanzenforschung under direction of Dr. Hans Stubbe.

Establishment of the facility was decided by the senate of the Kaiser Wilhelm Gesellschaft (KWG] in May 1939. But the institute was founded as late as only on 1 April 1943 in Tuttenhof near Vienna not least with Himmler's support. Work directly related to war was presumably the main task of a facility founded at a time when victory was no longer an ally of Germany's army. At least three of the researchers who have been involved under Stubbe in the research of weeds, Hansjörg Eichler, Werner Rothmaler, and Otto Schwarz, had been released from the army for that purpose at a time when Hitler had ordered to comb out the very last man able to serve in the army.

---

[6] in contrast to its director who, after the war, was appointed Professor of Philosophy at the Free University in West-Berlin.

Two research projects were classified with high priority grades. One of them was mutation research not coordinated by Blome. The field covered by the second high priority project and its coordinator are still unknown.

There are several indications that the experiments performed by Stubbe and his facility were coordinated by Blome.

The experiments suffered under the progress of the war, especially under bomb attacks and finally had to be stopped. Nevertheless the facility can be regarded as a BTW facility because the work carried out with weeds clearly had offensive intentions. After the war the facility was converted and translocated; as "Institut für Genetik und Kulturpflanzenforschung" it became the leading East German genetics institute.

4. Conclusions

It is too simplistic to designate the termination of activities of a BTW facility as conversion. Conversion is one of several ways in which prohibited activities of such facilities may be terminated. This can be illustrated by the fate of former German BTW facilities and facilities involved in BW defense activities after WWII.

Only a few German institutes have been involved in offensive BTW activities. With one exception (the Kliewe laboratory) there was no German facility solely involved in BTW activities, be they performed with offensive or defensive intentions. Even the Zentralinstitut für Krebsforschung Posen-Nesselstedt, which was *planned* to become *the* major German BTW facility, was designed as dual-use institute. "Cancer research" was not just a cover name but one of the two main tasks of the facility. That part of the Central Institute for Cancer Research destined for BTW activities was not completed by the end of the war. It is unknown, whether those parts of the institute already ready for use and its equipment were totally destroyed as consequence of the war events and whether some of the equipment was dismanteld by Soviet authorities.

The only one facility fully involved in BTW activities and operating during 1941-45 was a small department of the Institute for General and Military Hygiene of the Military Medical Academy in Berlin, the "Kliewe laboratory".

Two of the BTW facilities, the Kliewe laboratory and the Institute of Entomology of the Waffen-SS and Police were more or less destroyed and closed. Whether their equipment was dismanteld by Soviet and US authorities is not known so far.

Most of the BTW facilities were converted after the war: the Potato Beetle Research Station, the test field Raubkammer, the Kaiser Wilhelm Institute for Horticulture Research, and the State Research Institute Insel Riems. In all cases these facilities continued to perform civilian R & D and production activities which had also been carried out during the war simultaneously with BTW activities.

One institute, the Institute for Microbiology of the Army Sachsenburg, which was claimed to be a BTW facility at the Nuremberg trial, was involved not in BTW activities but in BTW defence and civilian health care activities only. Nevertheless the institute was closed and dismantled by the end of the war.

None of the German converted BTW facilities reverted within the 50 years which passed since the German defeat although they continued to work in the same field as during the time during their involvement in BTW activities. This demonstrates that provided there is a clear political decision by the national authorities to completely renounce the development and production of biological and toxin weapons, then there is little risk as to the danger of reversion of converted facilities that operate in the BTW defense field. One has to consider, however, that there is a depositary of the BTW which has violated the convention for some two decades (see Leitenberg, this volume). In such and similar cases the termination of BTW facilities must be performed in such a way as to exclude any continuation of BTW activities or any reversion.

5. Acknowledgements

The author acknowledges with gratitude the helpful comments he received from Professors John Ellis van Courtland Moon and Graham Pearson. A generous grant from the Volkswagen Foundation allowed the author to carry out his research. Dr. Friedrich Hansen kindly provided Ciesluk's report [28], translated into German by Eva Langowska.

6. References

1. *Protocol for the Prohibition of the Use in War of Asphyxiating, Poisonous or Other Gases, and of Bacteriological Methods of Warfare* (1925), reprinted in Geissler [note 15], p. 131.
2. *Convention on the Prohibition of the Development, Production and Stockpiling of Bacteriological (Biological) and Toxin Weapons and on Their Destruction*, reprinted in E. Geissler (ed.) 1986, *Biological and Toxin Weapons Today*, Oxford University Press, Oxford, p. 131.
3. Hedén, C-G. (1967) Defences against biological warfare. *Annual Review of Microbiology* **21**, 639-677, see especially p. 665.
4. Resolution adopted at the closing Plenary Session of the 10th International Congress for Microbiology, Mexico, 14 August 1970- Reprinted in *Bulletin of Peace Proposals* (1991), no. 2, p. 116.
5. Hedén, C.-G. (1971) BW-disarmament: Status and prognosis. In Executive Council of Gewerkschaft Wissenschaft for the World Federation of Scientific Workers, *ABC Weapons, Disarmament and the Responsibility of Scientists*, Berlin, 94-100, see especially p. 97.
6. Geissler, E. (ed.) (1986) *Biological and Toxin Weapons Today*. Oxford University Press, Oxford.
7. Geissler, E. (ed.) (1990) *Strengthening the Biological Weapons Convention by Confidence-Building Measures*, SIPRI Chemical and Biological Warfare Studies 10, Oxford University Press.
8. USA (1986), Statement by Ambassador Donald Lowitz, Head of the delegation of the United States of America to the Second Review Conference of the Parties to the Biological and Toxin Weapons Convention, 9 September.
9. Geissler, E. (1992) Vaccines for Peace: an international program of development and use of vaccines against dual-threat agents. *Politics and the Life Sciences* **11**, see especially p. 236.
10. Geissler, E. (1994) Arms control, health care and technology transfer under the Vaccines for Peace programme. In E. Geissler and J.P. Woodall (eds.) *Control of Dual-Threat Agents: The Vaccines for Peace Programme*, Oxford University Press, Oxford, pp. 10-39, see especially p. 33.
11. Leitenberg, M. (1994) The conversion of biological warfare research and development facilities to peaceful uses, in E. Geissler and J.P. Woodall (note 19), pp. 77-105.

12. Comptroller General of the United States (1972) *Problems Associated with Converting Defense Research Facilities to Meet Different Needs: The Case for Fort Detrick.* Report to the Congress. B-160140, US Government Printing Office, Washington, 16 Feb., quoted by Leitenberg (note 20).
13. Lajoie, R. (1997), quoted in *The CBW Conventions Bulletin*, no. 36, p. 13.
14. Geißler, E. (1998) *Biologische Waffen - nicht in Hitlers Arsenalen. Biologische und Toxin-Kampfmittel in Deutschland 1915-1945*, LIT-Verlag, Münster.
15. Geissler, E. (1997) Anwendung von Seuchenmitteln gegen Menschen nicht erwünscht, *Militärgeschichtliche Mitteilungen*, **56**, 107-55.
16. Wheelis, M. (1998) Biological sabotage in the First World War, in E. Geissler and J.E. van Courtland Moon (eds.), *The Genesis of Germ Warfare: The Development and Use of Biological and Toxin Weapons from the Middle Ages to the End of World War II.* SIPRI Chemical & Biological Warfare Studies **17**, Oxford University Press (in press).
17. Committee of Imperial Defence. Sub-Committee on Bacteriological Warfare (1937) Germany. Bacteriological Warfare. Note by Joint Secretary. 25 October. Public Record Office (PRO) WO188/650 PX010754.
18. Committee of Imperial Defence (1937) Chemical Warfare. 23 December. PRO WO188/650 XP010754, p. 15.
19. Committee of Imperial Defence. Sub-Committee on Emergency Public Health Laboratory Services (1939) Bacteriological warfare activities in Germany (Anthrax). Note by the Joint Secretary. CBW 45. 21 August. PRO WO188/650 XP010754.
20. Klee, E. (1997) *Auschwitz, die NS-Medizin und ihre Opfer*, 3. Aufl., S. Fischer Verlag, Frankfurt/Main.
21. Kliewe, H. (1941) Report submitted to Wa Prüf 9 [CW Defence Section of the German Army Ordnancy), 15 July. Available only in part as English translation in Alsos Mission (1945) *A Review of German Activities in the Field of Biological Warfare*, 12 September. Report No. B-C-H-H/305, MIS, War Department, Washington, D.C. National Archives Washington, DC, RG 112 C-4. pp. 56-57.
22. Fischer, H. (1985) *Die militärärztliche Akademie 1934-1945*, Biblio Verlag, Osnabrück.
23. Lebus (1943) Aktennotiz. Geheim. 30 October. Bundesarchiv Koblenz R2 / 11654.
24. Schreiber, W. (1945) Schreiben an die Regierung der Sowjetunion, Moskau, 10. April. Bundearchiv Militärarchiv, Freiburg/Brsg., N54/34.
25. Hamerla, K. (1989) *Geschichte, Geschichten und Bilder aus der Raubkammer*, Wehrwissenschaftliche Dienststelle der Bundeswehr für ABC-Schutz, Munster.
26. Federal Republic of Germany (1997), Annual Report of Confidence Building Measures. 23 April. UN Report CDA/BWC/1997/CBM, pp.315-27, see especially p. 318.
27. Hansen, F. (1993) *Biologische Kriegsführung im Dritten Reich.* Campus-Verlag, Frankfurt; New York, p. 147.
28. Ciesluk, H. (1951), Documents 277-283 from file [PH-W) Bd.304 abc, archives of the Polish Main Commission for persecution of Hitler crimes. Public prosecutor of district Posen, II.Prot.57/51, pp. 1-8.
29. Kater, M.H. (1972) *Das "Ahnenerbe" der SS 1935-45.* Deutsche Verlagsanstalt, Stuttgart.

CHALLENGES OF BW CONTROL AND DEFENSE DURING ARMS REDUCTION

GÁBOR FALUDI
Medical Corps of the Hungarian Defence Forces

1. Introduction

In this paper military aspects of former and current problems of biological warfare in Hungary will be illustrated.First I will sketch the changes of Hungarian military attitudes in comparison to those of the previous periods on biological weapons. Then I should like to summarize the salient trends of the arms-reduction process of the Hungarian Defense Forces (HDF). Finally I shall discuss some of the challenges associated with the transition to a lower level of biological defense.

In March 1996 I was invited as a military epidemiologist to participate in a NATO Advanced Research Workshop in Budapest, organized by the Budapest Center for Arms Control and Democratic Control. The title of the conference was "The Technology of Biological Arms Control and Disarmament". I was truly shocked to realize the degree of difference between Hungarian contemporary principles and those of the NATO as to the status of defense against biological weapons. This surprise directed me again to compile both older and more recent data and to investigate available documents. The "club" of military experts of biological weapons is not at all overcrowded in Hungary. Traditionally, members of two army services, the NBC defense and medical officers, dealt with problems of biological weapons. But, as an unwanted side-effect of the BTWC, the greater part of the Hungarian military and medical societies (in the latter case both the civil and military) became quite cavalier to the whole topic of biological weaponization. As a young regiment surgeon frequently heard from my older, high-ranking colleagues that "in a country possessing a well organized public health service in addition to a good curative medical system, the application of biological weapons has no real value or benefit". Indeed less than ten unclassified military and civil medical books, booklets and reports on the topic have been edited since the early fifties in Hungary.

2. Changes in Hungarian military BW attitudes

Fortunately, I was able to find an excellent retired NBC officer colleague to help me collect many old, formerly secret, and hidden documents containing information

E. Geissler et al. (eds.), Conversion of Former BTW Facilities, 67-72.
© 1998 *Kluwer Academic Publishers. Printed in the Netherlands.*

completely forgotten since the early 50's. It is on the basis of these that I can inform you about the early events of the Hungarian biological warfare program.

Theoretically, the history of Hungarian military initiatives on biological weapons can be divided into three different periods: a) the years between 1936 and 1944, b) the era of the Warsaw Pact from 1945 until 1989, and c) the solitary period of "standing alone",i.e. the period of preparation for Euro-Atlantic integration.

2.1. THE HISTORICAL PERIOD BETWEEN 1936 AND 1944

The Headquarters of the Hungarian Royal Defense Forces, with authorization from the Highest Defense Council in 1936 (the same year Porton Down was founded), discussed and decided upon the future of research and possible application of biological weapons in wartime. The Chief of HQ was ordered to organize a secret project with a team of experts and to create a small institute in Budapest with on the basis of reconnaissance obtained from neighbouring states about similar activities. The team leader was a medical bacteriologist, colonel Dezső Bartos, who organized and planned a Health Control Station in the inner territory of the Artillery Equipment Warehouse in Timot street in Budapest. The team consisted of: one leading bacteriologist-epidemiologist, one medical bacteriologist, one veterinarian parasitologist, one chemical engineer and two laboratory technicians. The institute was situated in a closed area, surrounded by a five-meter (17 feet) high rampart with a three-meter (ten feet) wide wired-off promenade. Work was initiated on 1. August 1938 in eight microbiological laboratories, one chemical laboratory, an animal house, and warehouse.

Until now only limited informations has been published about the activity of this Station. The first publication appeared in the "Handbook of Military History for NBC Officers", written by ret. col. P. Madaras in Hungarian [1]. The objectives of the research work were practically unknown even to the members of the Medical Service of the Hungarian Defense Forces (HDF). The source of this publication was a report that Bartos had prepared for the Minister of Defense in 1955 after his return from the USSR, where he had been inferred as a prisoner of war [2]. The 73 page secret report summarized the complete activity of Health Control Station.

Scientific research began at the Station in 1938 for both offensive and defensive purposes.The offensive aims were the following: Bartos laid down the basic principles of biological war. He identified the main possible targets as: the human population, domestic animals and agricultural plants. The practical military aims of application were investigated at tactical levels against troops, logistic and transport centers, as well as strategic initiatives against crowded cities. He proposed to evaluate geographical variations in the planning of biological operations.

Bartos' team determined three military methods of biological attack: the first was via airplane and artillary . The second method was by the employment of secret agents, and the third was to contaminate evacuated territory and wells upon strategic retreat. The team also tested the application of explosive wet and dry aerosols in field experiments using 1- to 50-kilogram glass bombs with stained indicator strains. This work aimed to explore the optimal meteorological conditions of attack, the effective concentration of germs per surface unit, and the most suitable species of bacteria, namely, Bacillus anthracis, Clostridium perfringens, Salmonella paratyphi and Shigella dysenteriae. The Bartos team carried out experiments with bacterioligally loaded from 9 mm pistols and guns. Their results showed that the pathogenic strains were able to survive the mechanic and thermal shock of explosive acceleration and that they could conclusively cause serious infective injuries.

Another line of their investigation was a phenotipically enhanced virulent Salmonella paratyphi strain selected by serial passage and tested on small vaccinated laboratory animal communities. The Bartos team further investigated the production of experimental tools for the sabotage of secret agents: infected toothpastes, candees and so on. In those years the long-time storage of pathogenic strains was a great problem. Bartos et al. nevertheless succeeded in developing a method of preservation by keeping the strains in pre-sterilized maize-seeds for extended periods at room temperature without loss of viability and virulence.

For defensive aims Bartos recommended several different methods. He suggested the introduction of an early warning system against the recognized infectious diseases. Several rules of collective defense were elaborated for a strictly guarded water supply systems and for the organization of an Animal Health System and Plant Defense Service. Unfortunately, half of the 73 pages of Bartos's report consisted of those flowery Stalinist rhetoric expressions that were characteristic for that time in eastern countries and perhaps were exigenicies of survival . In any case, the rest of the report indicated that before and during the World War II biological weapons were " in the air".

The BW-Station attempted to build up official international connections for research activities in association with similar institutions in Germany and in Italy. As we know from Geissler [3] and Hansen [4], biological weapon developments were prohibited at that time in Germany by Hitler so that such attempts could not obtain governmental support. Bartos attributed this to lack of confidence in his report.

In contrast to Germany, the Italian connection was more successful. The partner institute was situated in a military hospital in Rome. It was a large one, directed by lt. col. Prof. Raitano. Delegations of the two countries visited each other and exchanged information on their research activities.

The financial backing of the Hungarian project was very impressive: the cost of foundation was 200.000 P (former Hungarian currency) corresponding at that time to US $ 100,000 (1997 rate of conversion). Nevertheless, the monthly budget was only 1500 P currency units. The institute was built up within two years. During its six-year existence achievements comparable to other international efforts were produced. I don't wish to justify the quality of the report, but it seems to be comparable to other of that time. The project suddenly ended on the 6 April1944 as a consequence of a successful air-raid by the Allied Air Forces.

The experimental scale production ended before the introduction of large-scale technology. All the bacterial strains, records, materials and methods were destroyed together with the early offensive plans of the military administration of Horthy's

Defense Forces, and bombs otherwise demolished the total Station. The founder, Colonel Bartos, was sentenced to a prison of war for years in the USSR. Nevertheless, the story has also a happy end. After the completion of his report Bartos was fell back again into lucky anonymity and oblivion, and worked as a family doctor in Budapest until his death. Nobody knew of his significant military-scientific achievements; even his daughter claimed ignorance of such information. His report also slept in anonymity nearly forgotten in a secret file for decades before its scientific evaluation.

2.2. THE ERA OF THE WARSAW PACT UNTIL 1989

In this period ,the whole life and activity of our army was generally determinated by its status in the bipolar world. Among other weapons of mass destruction biological weapons maintained a subordinate significance and stayed in the shadow of nuclear and chemical weapons in Hungary. After the ratification process of the BTWC in 1975, biological defense lost even this residual importance. The NBC service organized methods and devices of individual and collective defense and the medical service dealt with epidemiological and micobiological problems of BW as a part of medical service, e.g. sampling, detection and identification of microorganisms, by rapid diagnostic procedures. The NBC and medical services worked nearly in isolation from each other. The activities were corresponding to the rules of Warsaw Pact. The medical service developed a mobile anti-epidemic squad in the sixties similar to those of the Soviet, Yugoslavian and East-German Armies.

The last of the defensive medical developments were carried out at the beginning of the eighties. A new generation of mobile laboratories in containers was developed to replace the so-called "fixed-on-truck" varieties. The anti-epidemic unit was formed from these mobile containers, which had no defined task except to detect agents of biological weapons under the conditions of traditional warfare. The unit was designed to solve problems of sampling, detection, identification and decontamination of troops and quaranteenization. The containers were produced of metal,equipped with tents and independent energy-supply systems, so was to continue investigations alone or connected to a military field-hospital. Critically reviewed, these laboratories had several severe disadvantages. Equipment from non-Warsaw manufacturers were prohibited, and so the containers could support routine diagnostic work only for conventional medical teams and diseases. Such developments had been introduced almost twenty years before, and the whole system could thus be considered to be obsolate.

In summary, the whole period was dominated by a defensive orientation and a modest indifference towards biological weapons. The defensive systems against the mass destruction weapons were elaborated at an acceptable average level. Sometimes one could receive troubling news about secret USSR activities such as ' The Sverdlovsk Accident' from popular news. It later became possible to find scientific articles written by M. Leitenberg [5-7] and M. Meselson [8-9] dealing with that event. Otherwise we lived substantially uninformed.

2.3. THE PERIOD OF "STANDING ALONE" AND PREPARING FOR EURO-ATLANTIC INTEGRATION

This newest era began with great euphoria. The old rules, and disputable military doctrines were thrown out upon leaving the WP. The world changed profoundly, and military principles have also changed in the light of possible warfare and local conflicts in the current European security environment. The theories of biodefense have had to adapt gradually to NATO standards.

3. The influence of arms reduction

Governmental reorganization to a pluralistic parliamentary democracy founded on a new constitution brought increased considerations of human rights and preparation for a free market economy. Crisis-management resulted in the reduction of both military budget and staffing. Let me cite the words of George Keleti, our Minister of Defense: "We are going to continue the execution of concepts for the transformation of the Military and the establishment of a smaller, but more efficient army. Restructuring involves the large-scale reduction of peace-time authorized military strength." [10] The new National Defense Policy is founded on decisions of the Hungarian Parliament, which has defined both medium and long-term directions of modernization. The declared aims include Euro-Atlantic integration, the realization of democratic and civilian control over the military and the gradual modernization of armed forces corresponding to the country's economic capabilities.

Significantly these forced army reductions might also diminish the level of defense prepredness against biological weapons. Theoretically, every process of reorganization involves a certain degree at temporal lability and vulnerability.

4. Challenges for BW control and defense

Fundamental governmental changes have brought new challenges at both state and military levels in Hungary. Our country has become totally open; according to the data of Statistical Yearbooks nearly 40 million foreign people entered the country each year between 1990 and 1995, and about 13 million Hungarian citizens traveled abroad. These are enormous numbers in comparison to the years before '89. This simple fact bears directly upon the new risk of emerging infections in Hungary. This year for example we diagnosed our first cholera case in 70 years. In addition the great diphtheria outbreak at our eastern borders, the West Nile virus epidemic near Bucharest and the Coxsackie B outbreak in Cyprus, along with elevated number of refugees from many countries, increased the risk of the spread of imported infections. These events oblige us to maintain a strong surveillance system against infectious diseases. The uncontrollable mass-movement of people is also a source of danger as outlined by Mr. Eifried's information on bioterrorism [11, 12].

Another risk arises from the great quantity -- the approximate number is one million according to the Central statistic office -- of small private economic enterprises

created by the free market economy. In this magnitude it is practically impossible to implement regular checking of all individual activities. While these new interprises opportunities for the economical development of the country in the future, they also create many dangerous opportunities for extremists in production and risks for public health that challenge present day biodefense.
The declared strategy of HDF -- the integration into NATO, and the participation in SFOR and many other multinational peacekeeping actions in numerous areas of the world (organized under the auspices of UN or the Organization for Security and Cooperation in Europe) -- also could bring further new problems to Hungary. We could, for example, become possible target for bioterroristic revenge operations.

Unfortunately, in contrast to the PROCEID recommendations on enhanced peaceful international co-operation in molecular medicine and biotechnology (Hunger, this volume), the underdeveloped level of medical-scientific cooperation in the diagnostic work, due to the lack of financial support and resources, has weakened the level of research activities. This situation reduces the possibility of rapid identification of the etiology of possible bioterror device and epidemiological hazards.

I personally believe that the mentioned risk factors temporarily affecting the biodefensive preparedness of the army should be regarded as a sign of higher threatening. I think these processes might also influence our mutual security in wider relations. According to the BTWC , significant diplomatic efforts are to be undertaken in order to enhance global BW defense at the international governmental level. In spite of this, a modest form of impaired security may reappear in our country. In order to respond to this challenge, it is reasonable to sustain the defense capacity at a minimal acceptable level with international efforts to reduce the danger of bio-terroristic actions.

5. References

1. Madaras, P. (1976). Hadtörténelmi kézikönyv vegyvédelmi tisztek számára. I. vol. A tömegpusztító fegyverek fejlődéstörténete. Magyar Néphadsereg Vegyvédelmi Főnökség Budapest, 1976.
2. Bartos, D. (1955), secret manuscript, unpublished.
3. Geißler, E. (1998) *Biologische Waffen - nicht in Hitlers Arsenalen. Biologische und Toxin-Kampfmittel in Deutschland von 1915 bis 1945.* LIT-Verlag, Münster - Hamburg.
4. Hansen, F. (1993) *Biologische Kriegsführung im Dritten Reich.* Campus-Verlag, Frankfurt - New York.
5. Leitenberg, M. (1991) Soviet activities related to biological weapons. *Arms Control* **12**, No. 2, 161-190.
6. Leitenberg, M. (1993) The biological weapons program of the former Soviet Union. *Biologicals* **21**, No. 3, 187-191.
7. Leitenberg, M. (1994) The conversion of biological warfare research and development facilities to peaceful uses, in E. Geissler and J.P. Woodall (eds.), *Control of Dual-Threat Agents: The Vaccines for Peace Programme*, Oxford University Press, Oxford, pp. 77-105.
8. Meselson, M. (1993) New evidence on the 1979 Soviet anthrax epidemic', *ASA Newsletter* **93-2**, pp. 1, 5.
9. Meselson, M. et al. (1994) The Sverdlovsk anthrax outbreak of 1979. *Science* **266**, 1202-08.
10. Keleti, G. (1996) Foreword in National Defence '96. *MoD Press Department*, Budapest.
11. Eifried, G. (1996) Responding to biological weapons terrorism. Paper presented at the NATO Advanced Research Workshop "The Technology of Biological Arms Control and Disarmament", Budapest, 28-30 March.
12. Eifried, G. (1996) On countering the threat of chemical and biological terrorism, *Politics and the Life Sciences* **15**, 199-201.

CONVERSION OF PAST BIOLOGICAL WEAPONS FACILITIES : LESSONS FROM WESTERN CONVERSION

GRAHAM S PEARSON
Visiting Professor of International Security
Department of Peace Studies University of Bradford
Bradford, West Yorkshire BD 7 1 DP United Kingdom

1. Introduction

The NATO Advanced Research Workshop (ARW) on the Conversion of Former Biological Weapons Facilities is timely as it enables us to review the nature of the former United Kingdom (UK) biological weapons (BW) programme which was abandoned in the late 1950s and to examine the lessons that can be learnt from the conversion of the British programme. In this article, these lessons are compared to those which can be learnt from the experience of UNSCOM in destroying the Iraqi BW programme so that points can be developed that are applicable when considering the conversion of the Russian BW facilities.

It should be noted that Article II of the Biological and Toxin Weapons Convention (BTWC) [1] which was opened for signature in 1972 and entered into force in 1975 makes no mention of BW facilities. Article II states that:

> *Each State Party to this Convention undertake to destroy, or to divert to peaceful purposes, as soon as possible but not later than nine months after entry into force of the Convention, all agents, toxins, weapons, equipment and means of delivery specified in Article I of the Convention, which are in its possession or under its jurisdiction or control. In implementing the provisions of this article all necessary safety precautions shall be observed to protect populations and the environment.*

This is in marked contrast to the Chemical Weapons Convention (CWC) [2] which opened for signature in 1993 and entered into force on 29 April 1997 which in Article V addresses particular attention to chemical weapon (CW) production facilities and requires their destruction "as soon as possible". Their conversion can only be exceptionally approved and, if so, is then subject to a more stringent verification regime.

The difference between the two Conventions reflects not only the different decades in which they were negotiated but also the different nature of CW and BW. As several CW agents -- such as mustard and the nerve agents -- have little if any civil or other permitted use, there are no alternative uses to which past CW production facilities can be put. On the other hand, BW agents are the same pathogens that occur naturally as

E. Geissler et al. (eds.), Conversion of Former BTW Facilities, 73-83.
© 1998 *Kluwer Academic Publishers. Printed in the Netherlands.*

disease -- after all biological warfare is the use of disease as a weapon of war. Consequently BW facilities can be used for peaceful permitted purposes to counter disease and so benefit public health and the environment. However, this dual purpose nature that is integral to microbiology means that whilst facilities can be converted from BW to peaceful purposes, care needs to be taken to ensure that there is no reversion of such facilities from permitted to prohibited purposes. The need to ensure that there is no reversion is an important point in considering the conversion of BW facilities and the monitoring and verification regime to assure this.

2. UK Past BW Programme

The elements of a BW programme which need to be considered in addressing conversion are the facilities, the activities and the personnel together with the organizational and political chain of command. A biological weapon can conveniently be regarded as being made up of the biological agent -- the microbiological organism or toxin which causes the desired effect against the target population whether that be humans, animals or plants -- and the dispersion system that delivers the agent to the target. The facilities associated with this are the research, development, production and storage capabilities. The organizational and political structure are an important part in the controls that this provides to the activities of the personnel engaged in the programme.

Much has been written about the past UK BW programme [3] so only a brief outline is provided here. Essentially concern arose in the 1930s as to whether there might be German interest in BW. This resulted after the outbreak of World War II in the establishment under great secrecy of what was known as Biology Department Porton (BDP) as a largely autonomous organisation within what was then the Chemical Defence Experimental Station. The names of both of these establishments have changed through the years since then as indicated in the table.

TABLE 1. UK Chemical & Biological Facilities

Year	Chemical	Biological
1930	Chemical Defence Experimental Station	
1940		Biology Department Porton
1946		Microbiological Research Department
1948	Chemical Defence Experimental Establishment	
1957		Microbiological Research Establishment
1970	Chemical Defence Establishment	
1979		Centre for Applied Microbiology & Research
1991	Chemical and Biological Defence Establishment	
1995	Defence Evaluation & Research Establishment	

The initial role of the Biology Department Porton (BDP) was to determine whether BW was indeed feasible, and, if it was feasible, to determine how the UK might retaliate in kind should British Forces be attacked by biological weapons. Such a retaliatory capability was in keeping with the 1925 Geneva Protocol which the UK had signed with a reservation that would allow the UK to respond in kind were the UK to be attacked by BW (or by CW). The work of BDP was carried out by experiments at Porton and by trials in 1942 and 1943 at Gruinard Island, a small island just off the north-west coast of Scotland. The feasibility of BW was demonstrated and work then commenced on the retaliatory capability.

The initial and, in the event, only British retaliatory BW capability was an anti-livestock weapons consisting of cattle cakes made of ground linseed meal (about 2.5 cm in diameter and 1 cm thick) which contained a lethal ingestion dose of anthrax spores. Work at BDP had shown that cattle would seek out and eat these cattle cakes and consequently a stockpile was created by April 1943 of 5,000,000 anthrax cattle cakes. If the UK or British Forces had been attacked by BW these cattle cakes would have been disseminated through the flare chutes of British bombers flying over Germany. The aim being that cattle would eat these cakes, catch anthrax and die so affecting the food chain and thus the German war effort. It is worth noting this production was carried out by one technician from BDP, 15 ladies from a Bristol soap factory who were employed at Porton during the war for ad hoc production tasks, one laboratory assistant, two labourers and one boy to assist. This team was not immunized and wore no respiratory protection; a simple plate of glass was present to prevent drops splashing from the anthrax spores dispensing device; none of the workers caught anthrax. The point to be noted is that high containment was unknown in the 1940s. Careful attention to cleanliness in the laboratory sufficed. Consequently, high containment is not an essential prerequisite for a BW capability.

The work on Gruinard Island was to determine whether sufficient quantities of anthrax spores could be produced in the appropriate range of particle sizes to infect and cause casualties in animals located downwind -- and thus presumably in men. The trials used small 4lb bomblets filled with anthrax spores which were exploded upwind of tethered sheep. These showed that sheep were infected and died from inhalation of the anthrax spores. The 4lb bomblets were envisaged as being delivered by 500lb cluster bombs which would have contained over 100 of the 4lb bomblets. There was close Allied cooperation between the UK, USA and Canada [4] and plans were made to produce anthrax spores in the USA and for field trials with cluster bombs in Canada. In the event fulfillment of the Allied plans was halted by the end of World War II, before the anthrax production plant in the USA had produced any anthrax spores.

Following the end of World War II, the anthrax cattle cakes were incinerated. Gruinard Island was visited; samples showed that there were still viable anthrax spores on the island. It was therefore decided in 1946 to purchase the island for £500 with the understanding that when it was "safe for man and beast" the island would be offered back to the owner for the same sum. Also in 1946, the name of BDP changed to Microbiological Research Department (MRD) still colocated within the then CDES and still an autonomous unit. The success of the Gruinard Island trials and subsequent research by the UK, US and Canada convinced the UK Government that further study of biological warfare was essential.

Although the development of a UK nuclear weapon capability had had a significant effect on UK military thinking, biological warfare was perceived as being of great significance. Consequently, in austerity Britain, a new £2.5 million building on a site adjacent to CDES commenced construction in 1948 and was occupied by MRD in 1951. The objective of MRD was for detailed and long range study of the processes underlying biological warfare and defence. Further trials took place in the post war years starting in 1948 which were carried out at sea either off the Bahamas or in Scottish waters. These trials showed that the strategic potentials of biological warfare were considerable and that use of BW against the UK or UK forces presented a significant hazard.

In 1957 MRD became a separate establishment, the Microbiological Research Establishment (MRE). The late 1950s also saw a UK decision to abandon all aims to develop an offensive biological warfare capability. As very few staff at MRD/MRE had been involved with offensive matters since World War II, this decision had little effect at MRE as at both MRD & MRE the emphasis had been on research rather than development and no specifically offensive research was in progress.

3. UK Conversion

As already noted, MRD had become MRE in 1957 and was already engaged primarily in research activities. MRE was to have two clearly defined roles: to assess the hazard of biological warfare and to devise the means of protection of the UK Armed Forces. As Gradon Carter has eloquently described [5] "This was for many of the staff, a matter of fundamental research, where the link between topic and the role of the Establishment was tenuous." He noted that MRE attracted many leading scientists whose interest in defence was limited; they had been drawn to MRE by its reputation as a result of much open publication of its work and the superb facilities. Gradually, the majority of MRE scientists became involved in fundamental research work with civil applications that was not in reality aimed at biological defence.

By the mid 1960s, it was accepted that the interests of the Establishment should widen and the similarity of many problems whether resulting from use of biological warfare or from naturally occurring disease led to work being carried out on behalf of and funded by the Ministry of Health, the Medical Research Council, the Public Health Laboratory Service and industry. MRE became a centre for arbovirus research working on dangerous new diseases such as Ebola fever and Lassa fever. Increasingly the work was for civil purposes unconnected with biological defence. In 1968 there was a week of Open Days to dispel lingering suspicions that MRE was not engaged on solely defensive work.

In 1971 about a third of the funding of MRE was from civil sources and the civil and biological defence work was more nearly 50:50. The following year saw the signature of the Biological and Toxin Weapons Convention (BTWC) which entered into force in 1975. By the mid 1970s 90% of the work of the Establishment was being published in the open literature. However, the 1970s also saw the future existence of

MRE being questioned with the mid-1970s seeing a decision that a reduced programme of BW defence was acceptable and one that could be carried out by transferring a small team from MRE to the then Chemical Defence Establishment (CDE). The future of MRE was reviewed leading to the announcement that on 1 April 1979 the staff and facilities of MRE would be transferred to the Public Health Laboratory Service and become known as the Centre for Applied Microbiology and Research (CAMR). At the same time a small focus of expertise of about a dozen individuals to CDE where a Defence Microbiology Division was created to carry out biological defence studies.

The Centre for Applied Microbiology and Research (CAMR) in the 1980s developed its civil links and entered into a development and marketing agreement with a company known as Porton International. This led in the early 1990s to consideration by the UK Government whether to sell CAMR to a commercial company. After consideration, it was decided to retain CAMR within Government ownership and in 1994 CAMR left the Public Health Laboratory Service and came under the Microbiological Health Authority.

3.1 GRUINARD ISLAND

This had been visited annually to take samples of the soil and analyse these for the presence of anthrax spores which did not appear to diminish. However, in 1978 a sampling survey was carried out of the island which showed that a relatively small 3 acre area of the 550 acre island was contaminated with anthrax spores which were generally in the top few cms. During the next few years tests were carried out in which various decontaminants were evaluated both on the mainland and on the island. It was shown that a 5% solution of formaldehyde in sea water applied at the rate of 50 litres per square metre was effective in killing the anthrax spores and did not result in the terrain becoming bare infertile soil. Consequently, by 1985, the area that was contaminated was known and a technique was available to decontaminate the soil.

The Ministry of Defence decided that it would be inappropriate for the Ministry to both be the judge and the jury of the action proposed to be taken to clean up Gruinard Island so that it could be returned to civil use. Consequently, the Chief Scientific Adviser of the Ministry of Defence wrote to the President of the Royal Society inviting the Royal Society to appoint the Chairman of an Independent Advisory Group on Gruinard Island (IAGGI) who would then select some additional independent experts to review the proposals for decontaminating the island and then to monitor the decontamination and eventual return to civil use. The Royal Society appointed Professor William D P Stewart FRS (now Sir William Stewart), then the Boyd Baxter Professor of Biology at the University of Dundee to chair the IAGGI. The IAGGI reviewed all the work carried out thus far by the Ministry of Defence and accepted that a 5% solution of formaldehyde in sea-water would be effective in killing the anthrax spores. They recommended that some 10 acres of the island should be treated so as to ensure that not only the area known to be contaminated but also a surrounding area were treated. The contractors presented their approaches to achieve the 50 litres per square metre and Languard were selected to carry out the work.

The decontamination was carried out in the summer of 1986. Some three months later the area that had been decontaminated was resown with grass seed to encourage regrowth and soil samples were taken in duplicate down to the bedrock (generally about 50 cm below the surface); one set of samples was analysed at the Chemical and Biological Defence Establishment at Porton Down and the other set by the Central Veterinary Laboratory at Weybridge. The results of these analyses agreed well and showed that there was no trace of anthrax spores in the area treated except for a few isolated samples; some 9 out of 284 samples. These areas were retreated and subsequently resampled; no trace of anthrax spores remained after retreatment.

The following summer, at the request of the IAGGI, the Ministry of Defence arranged for a local farmer to place a flock of sheep on the island to demonstrate that the island could now be used in the same way as the land elsewhere in the area. The flock of sheep stayed on the island from May to October and flourished. They were checked monthly by a vet from the Scottish Veterinary Service. The IAGGI then prepared their final report in which they judged that Gruinard Island was now safe for man and beast. The Ministry of Defence returned the island to the heirs of its original owners in April 1990. [6]

4. Lessons for Conversion

Certain lessons can be drawn regarding conversion from the British experience. First, there was a very clear political decision in the late 1950s to abandon all work on an offensive BW programme. All subsequent work was aimed at providing effective BW defence. The cuts in defence funding in the 1960s and 1970s meant that the expertise in the Microbiological Research Establishment (MRE) sought funding from outside defence and increasingly took on and carried out civil work. There was increasing transparency about the work being carried out at the MRE through publication to an increasing extent in the open literature reaching about 90% and the holding of open days to inform those interested about the work of the Establishment and the contribution that it was increasingly making to public health in the UK. Increasingly opportunities have been sought to publish articles about both past and present activities. For several years, an annual meeting has been held by the Directors of the two Establishments at Porton Down -- the CBDE and CAMR -- with representatives from the seven Parish Councils which adjoin the boundary of the establishments. These representatives have been briefed about the work of the establishments and their plans for the future and they have the opportunity to ask questions about any aspects of the establishments. Such a meeting gives the local residents confidence that there are no dangers to them from the work of the establishments.

All of this is coupled with the public accountability and control that is common to all UK Government establishments. The Director is responsible and accountable to his superiors in London who in turn are responsible and accountable to their political masters, the Ministers who sit in Parliament and answer Parliamentary Questions about any and all aspects of the work carried out in the government owned establishments. Recent years have seen this linkage becoming even more direct with Directors becoming Chief Executives and directly answering Parliamentary Questions about the operations of the establishment whilst the Minister answers Parliamentary Questions relating to the policy about the establishment. In addition, government establishments have become increasingly subject to all the rules and regulations concerning their activities; the days in which government establishments had Crown immunity have long since gone. There is consequently public confidence that government establishments are only engaged on permitted activities and are monitored and inspected by various authorities such as the Health and Safety Executive.

5. Lessons from Iraq

There are also valuable lessons relating to conversion that can be learned from the experience of the United Nations Special Commission (UNSCOM) on Iraq. Security Council Resolution 687 (1991) [7] required that "Iraq shall unconditionally accept the destruction, removal or rendering harmless, under international supervision, of;

(a) All chemical and biological weapons and all stocks of agents and all related subsystems and components and all research, development, support and manufacturing facilities;

(b) All ballistic missiles with a range greater than 150 kilometres and related major parts, and repair and production facilities;"

and called for the provision to the Special Commission for destruction, removal or rendering harmless of all the items listed in subpara (a) above. The work of UNSCOM over the past 7 years has been remarkably successful in uncovering the Iraqi weapons of mass destruction programmes even though it has been clear throughout that Iraq has not taken the political decision to abandon the possession of such weapons and, indeed, has sought to hide as much as possible from UNSCOM.

This has been especially the case in respect of the Iraqi BW programme where as late as October 1997, the latest report [8] from UNSCOM to the Security Council, UNSCOM said that *"the September 1997 FFCD* [Full, Final and Complete Disclosure] ***fails to give a remotely credible account*** *of Iraq's biological warfare programme"* [Emphasis added]. Nevertheless, the information eventually revealed by Iraq to UNSCOM has resulted in the Al Hakam biological weapons facility being destroyed under UNSCOM supervision and other key equipment that did not appear to have legitimate uses has also been destroyed. Although UNSCOM has quite rightly repeatedly emphasised to Iraq that full information needs to be provided about past and present programmes so that UNSCOM can devise an ongoing monitoring and verification (OMV) programme to monitor the compliance of Iraq with its undertaking not to reacquire weapons of mass destruction, this information has not been forthcoming. Despite this, UNSCOM has succeeded in setting up an OMV programme under which some 90 biological sites were subject to regular monitoring and 893 items of equipment had been tagged. In the six month period up to October 1997, some 240 inspections had been undertaken.

The UNSCOM OMV system is comprehensive. It is based on declarations at six monthly intervals, the tagging of dual-purpose equipment and the monitoring of the activities at the 90 biological sites so that checks can be made as to what is being done. For each site, a comprehensive Protocol is generated which contains site information and the outcome of all previous inspections so that inspectors can readily identify changes in the activities being carried out. In addition, there is an extensive import/export monitoring mechanism in which any material or equipment on a lengthy detailed list has to be notified by the State considering exporting such material or equipment to the UNSCOM export/import unit; in addition, Iraq is also required to notify UNSCOM of any planned import. The ongoing monitoring and verification groups will then check that the imported material has indeed arrived at its declared destination, tag it if that is judged appropriate and also include it in the subsequent OMV programme.

The lessons from the UNSCOM experience for conversion are that a clear political decision to abandon the prohibited BW programme is an essential prerequisite. Full information needs to be provided about the past prohibited programme including details of the facilities, equipment, materials and personnel involved together with full information about the current use of these facilities, equipment, materials and personnel. Full transparency is vital if confidence is to be built.

6. Fissile Material Cutoff Treaty

Although negotiations for the Fissile Material Cutoff Treaty (FMCT) have yet to commence, several analysts have considered what elements are likely to be required for such a treaty. [9] The FMCT is envisaged as being very much complimentary to the Comprehensive Test Ban Treaty (CTBT) which has opened for signature but has yet to enter into force. The CTBT is seen as a qualitative cap to nuclear weapons as without testing, it will be very difficult to develop new improved nuclear weapons. FMCT is seen as the quantitative cap to nuclear weapons as it will stop the production of new fissile material. Both CTBT and FMCT will apply to both the Nuclear Weapon States and to Non-Nuclear Weapon States.

The requirements for verification of the future FMCT will be to ensure that:

* Facilities that have been shut down are not reused
  - this will require seals, monitors, inspections, and environmental samples
* Plants that continue to operate will be required to declare all procedures
  - this will require declarations, design verification, surveillance....
* Nothing is diverted from inventories, etc
  - this will require material accountancy, independent verification...

* Clandestine production is detected
  - this will require measures from 93 + 2 protocol

There is thus an interesting parallel between the requirements for verification of a future FMCT and the requirements for the monitoring of a converted BW facility. If the facility is shut down and is not destroyed, then it will be necessary to ensure that it is not unused. If the facility continues to be used for permitted purposes, there needs to be sufficient transparency to ensure that the activities are indeed permitted and that there is no diversion of materials and equipment for prohibited purposes. For confidence in all of this, there need to be declarations and for inspections to build confidence in the current activities.

7. Points for Russian Conversion

It is interesting to examine what points can be drawn from the above regarding the conversion of past Soviet Union BW facilities. It is clear that a principal requirement is for a clear political decision to abandon the past BW programme. Although President Yeltsin signed a decree [10] in April 1992 stating that the development and implementation of biological programmes in breach of the BTWC are not permitted on the territory of the Russian Federation, a trilateral UK/US/Russian initiative [11] started in September 1992 aimed at building confidence that the previous Soviet Union offensive BW programme which had continued until 1992 had indeed been terminated has not yet been successful. The 1995 report by the US Arms Control and Disarmament Agency (ACDA) to the US Congress submitted in July 1996 noted [12] that "Previous assessments of Russian compliance [with the BTWC] have highlighted the dichotomy between what appears to be the commitment from President Yeltsin and other members of the Russian leadership in attempting to resolve BWC issues and the continued involvement of 'old hands' in trilateral BW discussions and in what Russia describes as its defensive BW program." and goes on to conclude that "while there has been progress towards achieving the openness intended in the Joint Statement, the progress has not resolved all US concerns." The situation is described in the 1996 ACDA report submitted in mid-1997 by unchanged wording.

Further key requirements are that the past BW facilities are seen to be under clear political accountability and control. Those elements of the facilities that have no likely permitted use -- such as hardened facilities and explosive aerosol chambers -- should be destroyed thereby giving tangible evidence that past offensive capabilities are not being retained for possible subsequent misuse. The past facilities also need to be seen to be compliant with national rules and regulations for microbiological facilities and to be subjected to inspections by national authorities like other such facilities. There needs to be clear transparency about all activities -- both past and present -- with equal visibility about the people involved in the past and their current activities. All of this is vital to build confidence that the past BW programme has indeed been terminated and that the facilities and personnel are now engaged solely in permitted activities. Although conversion to BW defence can often be seen as an attractive first step in conversion, care needs to be taken that the BW defence programme is indeed credibly sized for the country concerned. Preferably, the past BW facilities and personnel should be converted to working on civil topics and microorganisms unrelated to those generally regarded as being potential BW agents. If funds are being sought to aid conversion from other countries, then all of the above is particularly important so that the country concerned can be confident that its financial aid cannot possibly be supporting the maintainance of a covert prohibited BW capability.

8. Conclusions

Consideration of the experience gained in the conversion of past UK BW facilities as well as consideration of the lessons to be learned from UNSCOM work in Iraq, demonstrate clearly the importance of the following points:

- Clear political decision to terminate the BW programme
- Facilities and personnel involved under firm political accountability and control
- Transparency about all activities, facilities and personnel both past and future
- Destruction of weapons only facilities
- Conversion to civil topics unrelated to BW defence builds greatest confidence

Given the attention paid to the conversion of past chemical weapons production facilities in the CWC, it is evident that past BW facilities will need to be addressed in the negotiations for a Protocol to strengthen the BTWC.

9. References

*This work was in part supported by a Research and Writing Grant from the John D. and Catherine T. MacArthur Foundation.

1. United Nations General Assembly, *Convention on the Prohibition of the Development, Production and Stockpiling of Bacteriological (Biological) and Toxin Weapons and on their Destruction,* Resolution 2826 (XXVI), 16 December 1971.
2. United Nations, *Convention on the Prohibition of the Development, Production, Stockpiling and Use of Chemical Weapons and on their Destruction,* United Nations 93-05070, 1993.
3. See for example, G B Carter, Porton Down: 75 Years of Chemical and Biological Research, Her Majesty's Office, London, 1992; G B Carter, Biological Warfare and Biological defence in the United Kingdom 1940-1979, RUSI Journal, 67-73, December 1992; Graham S Pearson, Biological Weapons: The British View, in Brad Roberts (ed), Biological Weapons: Weapons of the Future, The Center for Strategic and International Studies, Washington, DC, 1993; G B Carter & Graham S Pearson, British Biological Warfare and Biological Defence 1925-1945, in Erhard Geissler & John Moon, SIPRI, Oxford University Press, 1998(in press).
4. Gradon Carter and Graham S. Pearson, "North Atlantic Chemical and Biological Research Collaboration: 1916-1995, *Journal of Strategic Studies*, Vol. 19, No.1, March 1996, pp. 74-103.
5. G B Carter, Porton Down: 75 Years of Chemical and Biological Research, Her Majesty's Office, London, 1992
6. R J Manchee and W D P Stewart, *The decontamination of Gruinard Island,* Chemistry and Britain, July 1988, 690-1. Richard J Manchee et al, *Formaldehyde Solution Effectively Inactivates Spores of bacillus anthracis on the Scottish Island of Gruinard,* Applied and Environmental Microbiology, 60, No. 11, 4167-4171, November 1994.
7. United Nations Security Council, *Security Council Resolution establishing detailed measures for a cease fire, including deployment of a United Nations Observer Unit; arrangements for demarcating the Iraq-Kuwait borders; the removal or destruction of Iraqi weapons of mass destruction and measures to prevent their reconstitution, under the supervision of a Special Commission and Director General of the IAEA; and*

creation of a compensation fund to cover direct loss and damage resulting from Iraq's invasion of Kuwait, S/RES/687(1991) 3 April 1991.
8. United Nations Security Council, *Note by the Secretary-General*, S/1997/774, 6 October 1997.
9. Annette Schaper, *A Treaty on the Cutoff of Fissile Material for Nuclear Weapons -- What to Cover? How to Verify?*, Peace Research Institute Frankfurt, PRIF Report No 48, July 1997.
10. President Boris Yeltsin, Russian Federation Decree No 390, Moscow, 11 April 1992.
11. Richard Boucher, US Department of State, *Joint US/UK/Russian Statement on Biological Weapons*, Press Release, Washington Dc, 14 September 1992.
12. US Arms Control and Disarmament Agency, *Threat Control through Arms Control*, Annual Report to Congress 1995, 26 July 1996, pp66-67.

DEVELOPMENT OF BIOPHARMACEUTICAL MANUFACTURING AT FORT DETRICK, MARYLAND

GEORGE A. ROBERTSON, PH.D.
SAIC Frederick
PO Box B
Frederick, MD 21702-1201

1. Introduction

On Oct. 19, 1971, President Nixon founded the Frederick Cancer Research Facility as a way of demonstrating his commitment to eliminate the US offensive biological warfare program and to furthering his "War on Cancer". The following June, approximately 70 acres and 67 buildings were transferred to the National Institutes of Health. Since that time, almost every building has been renovated to accommodate new technologies and many several laboratories have been constructed. Five National Cancer Institute programs are based here, including the NCI AIDS Vaccine Research Program and its Supercomputer Program. Other Institutes have laboratories on campus, and the Frederick Cancer Research and Development Center, NCI-FCRDC (as it is now known) provides technical support to other NIH laboratories, government agencies and grantees worldwide. [1]

A significant amount of the support effort at the FCRDC is provided by contractors. Largest among these is Science Applications International Corporation (SAIC). SAIC employs over 1,000 scientific, technical and administrative personnel as the Operations and Technical Support (OTS) contractor of the FCRDC [2]. In 1993, the NCI directed the OTS contractor to establish an integrated manufacturing operation for the production of biopharmaceuticals for advanced preclinical and human clinical trials. Subsequently, the OTS contractor organized the program and designed, constructed, validated and currently operates Bldg. 459 for that purpose. The fermentation capability of Bldg. 472 was added to the program in 1996.

The purpose of this chapter is to describe how the concepts of U.S. Food and Drug Administration (FDA) Good Manufacturing Practices (GMP) and validation were applied to the qualification of these two buildings at Fort Detrick for the production of biopharmaceuticals for clinical and advanced preclinical studies for the National Cancer Institute.

E. Geissler et al. (eds.), Conversion of Former BTW Facilities 85-106.
© 1998 *Kluwer Academic Publishers. Printed in the Netherlands.*

2. Bldg. 459 - The Monoclonal Antibody / Recombinant Protein Production Facility

The building was constructed as a warehouse in the 1940s. It was never used for the production of any biological products. As the part of the NCI directive for the OTS contractor to establish a program to produce biologics, this building was completely renovated in 1994-1996. Bldg. 459 was a single story, rectangular structure 144 feet by 64 feet. The major constraint in developing this facility was the necessity to remain within the footprint of the original structure. This taken into account, the contractor was able to design over 8000 square feet of production areas to support the following functions: [3]

- Monoclonal antibodies - Quarantine and production cell culture laboratories, hollow-fiber bioreactors and freezer and liquid nitrogen storage for cell banks and products.
- Recombinant molecules - Microbial fermentation capability up to 80 liters and cell culture/bioreactor laboratories.
- BL-3 containment - For the production and filling of vaccines or gene therapy vectors using live agents.
- Aseptic operations / sterile filling - Class 10,000 and Class 100 cleanrooms for the purification and filling of protein, carbohydrate or peptide therapeutic agents. This includes support areas for washing, sterilization, buffer preparation and staging of equipment to be used in the sterile core.
- Research and Development - A fully-equipped biochemistry laboratory for the development and qualification of purification and analytical methodologies to support each clinical product.

3. Bldg. 472 - The Fermentation Pilot Plant

Built in 1952 to be used as a stimulant pilot plant to produce *Serratia marcesens* and *Bacillus subtilis* as stimulants. There is no facility biocontainment designed into the building architecture. A recovery and purification laboratory was added to the fermentation area and has been used by NCI for antibiotic and natural products production since 1976. Some of the utilities are currently in upgrade for recombinant protein production. Its capabilities include [4]:

- Fermentation at the 30 - 300 - 3000 gallon scale.
- Recovery and purification up to 2000 gallons, with the capability of using flammable or volatile solvents
- Seed development and cell banking.
- High capacity continuous flow centrifugation and filtration.
- Fermentation research and development with stand-alone fermentors with a capacity of up to 80 liters.

4. Current Good Manufacturing Practices (GMP)

In the United States, the Federal Food Drug and Cosmetic Act defines a system to ensure quality and consistent manufacture of pharmaceutical products. Known as "Good Manufacturing Practices" (GMP) or "Current Good Manufacturing Practices" (cGMP), the system can be described as "a set of scientifically sound methods, practices or principles that are implemented and documented during development and production to ensure consistent manufacture of safe, pure and potent products." [5, 6]. The Act further defines the term to include organization and personnel, buildings and facilities, equipment, control of components, drug product containers and closures, production and process controls, packaging and labeling control, holding and distribution, laboratory controls and records and reports. [5]. Generally, the GMPs describe "what" must be done, leaving latitude as to "how" the requirement is to be achieved.

4.1 PRINCIPLES ON WHICH GMPS ARE BASED

- Quality, safety effectiveness must be designed and built into the product.
- Quality cannot be inspected or tested into the product.
- Each step in the manufacturing process must be controlled to ensure final product meets quality and design specifications.
- Documentation evidences compliance with GMPs.

4.2 WHY FOLLOW GMPS?

This topic may seem trivial, as "It's the law!!" However, GMPs are scientifically based and have been accepted by regulatory authorities worldwide. As GMPs are being harmonized - US, EU, Japan [7], compliance will be necessary for international markets, regardless of local regulations. Put another way, although GMPs are primarily intended as regulations to protect the public, the result is to improve manufacturing consistency. The end result provides numerous benefits to the manufacturer, which include:

- Assurance of purity of drug products
- Assurance of consistency from batch to batch
- Provides a system to detect errors in manufacturing
- Reduces errors in manufacturing
- Increases productivity through greater consistency and less rework

4.3 TRANSLATION OF GMP REQUIREMENTS (PUT IT IN WRITING)

A hallmark of GMP is the requirement for documentation. This translates into a firm's relationship with its suppliers and contractors. The importance of well considered specifications cannot be overemphasized. Besides the obvious savings to the manufacturer in fewer change orders and more accurate bidding from contractors and suppliers, detailed specifications allow the quality assurance department and contract manager to be able to more precisely ascertain if the vendors are complying with the contract.

There are a number of ways to develop specifications. The most important source should be your in-house expertise. Consultants and the literature as well as meetings and seminars are frequently helpful, especially in obtaining other points of view. Vendors, as long as they do not recommend "lock out" specifications, are excellent, and often overlooked sources of technical information. These representatives are usually subject matter experts, and can make recommendations not only from their own technical knowledge, but also from their experiences with other customers. They can also provide references which will be an excellent "networking" tool.

A "Turnover Package" (TOP) is described as part of the Validation Master Plan Outline in section 9.5.3. In the TOP, a series of specification categories included. Where applicable, requirements for these data *should be included the purchase order.* The vendor should then provide information in sufficient detail to directly support the validation process. Experience has shown that a significant effort in detailing specifications and specifying the deliverables to the vendor results in a more efficient and far less stressful start-up and validation for the owner.

5. Validation

21CFR, Section 211.100 contains a general requirement that is cited for validation: "There shall be written procedures for production and process control to assure that the drug products have the identity, strength, quality and purity they purport or are represented to possess." Over the years, validation concepts have been applied to facility design, utility systems, operations, processes, assays and equipment design. For this paper, we shall consider how the facility as a whole can support a production process.

In the figure below, two types of validation are contrasted as they relate to supporting a process. *Prospective Validation* is most commonly encountered in the industry. After a new process (drug) has been selected for commercialization, scientists and engineers will design a facility which will provide the utility support and the environment for the manufacturing to successfully and consistently take place.

In the case of conversion, where a facility exists, the limitations on what prospective processes (or products) can be manufactured there are imposed by what can be validated. *Retrospective Validation* goes from the macro, the building and its systems, to the micro, the process equipment. In other words, which variables can be reliably controlled. A comparison is shown in the figure below.

TABLE 1. Comparison of prospective and retrospective validation.

PROSPECTIVE	RETROSPECTIVE
Key Process Variables	*Controllable Variables*
Process	Environment
Equipment	Utilities
Utilities	Equipment
Environment	Process

6. Organization for Validation

Validation is a multi- disciplinary effort which is not just a series of discrete steps or documents, but a continuous process which continues as long as the facility remains operational. In order to ensure the success of the process, an effective team approach is required. Our experience has shown that different specialties are required for each type of validation as well as during each stage.

6.1 PROSPECTIVE VALIDATION FOR NEW FACILITIES OR PROCESSES

In this case, a process or production goal had been defined, and the team must be constituted to provide the expertise to ensure the appropriate equipment and facilities are specified, constructed, qualified and operated for the consistent manufacture of a product. Here, scientists have the lead in advising the team as to the proper equipment and manufacturing environment (temperature, air quality) necessary to consistently produce the product. Once this has been determined, the architectural and engineering firm should begin to translate these requirements into design specifications and room criteria. Included in these specifications will be room and utility adjacencies based upon process flow, utility maintenance and economy. While specifications are being developed, the validation contractor (or in-house quality assurance unit) should be an active participant as advisors, as well as developing the Validation Master Plan, based upon the production process and the systems identified. At this point, the validation team is described as personnel or firms in the employ of the owner. Vendors and regulatory authorities can be equally important to the success of the validation effort.

Vendors can be valuable sources of specific product information. They will have detailed knowledge of their own products, but of their competitors' as well. Vendors should be able to provide you with references from current customers, who could also be useful sources of information about their experiences with similar processes. Vendors can also provide assistance in developing specifications, provided that they are not so specific as to exclude competitors. Early coordination with regulatory agencies is also helpful.. In the United States, the FDA actively encourages discussions early in the design and planning process. Although they cannot be part of the "team", comments from design presentations and reviews will provide insights into what they consider to be emerging quality trends and technologies which would impact design and construction decisions for your project. Once the quality criteria had been established, the engineering staff and project management and the quality unit ensure compliance during construction and commissioning [9, 10].

6.2 RETROSPECTIVE VALIDATION FOR EXISTING FACILITIES TO PRODUCE PHARMACEUTICALS TO GMP STANDARDS

Retrospective validation is more of an audit and a survey. A master validation plan would be developed by the quality unit in the context of the systems and design of the existing facility. It is probable that the facility has been operational for some period of time. As a result, there should be a number of highly qualified and knowledgeable employees available to join the validation team. These people would include: plant engineers, maintenance personnel, production technicians, contract services vendors and current equipment suppliers.

Drawing from this core of experts, the validation team initially gathers data, documents and drawings in order to determine the present state of the facility. The outline in section 9 highlights the major subject areas which the validation team will need to cover. If a QA/QC Department had not been established, an excellent first task would be to oversee the collection and organization of these documents. Organizations which did not have formal documentation programs in place may have to reconstruct histories. This is permissible, because regulatory authorities and potential customers will recognize that a retrospective validation is a "snapshot" of the facility in its current state.

The following documentation is required to support an retrospective validation:

- Overall building floor plans.
- "As built" - updated drawings.
- P&ID - piping and instrumentation diagrams for each system.
- Process flow diagrams for each proposed product.
- Floor plans with HVAC and pressure differential details.
- Equipment tagging and numbering schemes.
- Equipment schedules, to include all critical components, such as instrumentation, valves and materials of construction.
- Maintenance programs to include calibration protocols, schedules and records.
- Equipment cleaning and use logs, to include cleaning SOPs and acceptability criteria.
- Operating procedures.
- Equipment (systems) operational records.

Once the facility has been evaluated, decisions can be made as to what processes/ products can be manufactured. A firm might be advised that retrofitting can make a facility "GMP". This advice should be considered very skeptically, for many large, complex utility systems, quality begins at *design* and not simply in construction.

7. Components of biotech facility validation

7.1 ARCHITECTURE / DESIGN

There are two main considerations in designing a pharmaceutical production facility, to accommodate the process and to provide a smooth, segregated and safe manufacturing environment.

As discussed earlier, a manufacturing facility must provide space and utilities for efficient and validatable production. This not only includes the room environment, but also a facility which can be cleaned and maintained to support all phases of the operation. Most significantly is the concept of segregation of function and layout adjacencies. This can be as complex as a production line with inoculum preparation - fermentation - recovery - purification - final formulation - filling all in adjacent, one-way rooms or as simple as a facility performing one function only, such as a fermentation plant providing bulk product to be processed elsewhere. Also, within a facility, rooms should be designated, as much as possible, to one function or step in a process. This again could be as complex as a parenteral filling facility to an addition of a separate weighing room to a media formulation suite, to minimize component mix-up or cross-contamination.

Not only must a facility be efficiently designed on a macro scale, but room utilities, furniture and surfaces must be Spartan and cleanable to permit validatable room cleaning, either to separate products or production lots [8,11].

Important Design Considerations

- Support processes
- Segregate unit operations
- Layout adjacencies
- Minimize mix up / cross contamination
- Flow of material and personnel
- Separate clean / dirty
- Accommodate utilitiesHVA

7.2 HVAC (HEATING, VENTILATION AND AIR CONDITIONING)

- Supports the manufacturing process
- An “ingredient” (aseptic processing)
- Provides containment / isolation
- Protects products
- Protects people, environment
- Requires extensive monitoring and documentation

Biological products must, by their very nature, be produced in an environment which is free of extraneous contamination, therefore, the room environment is considered to be part of the manufacturing process. The room is the secondary barrier between the process and an uncontrolled environment. It provides a secondary containment for a toxic or biohazardous substance. Room conditions must be validated for temperature ranges, humidity and air quality appropriate for the operations being conducted. Accordingly, a well organized environmental monitoring program must be in place to document the HVAC performance in the following areas (as applicable): space temperature and humidity, pressure differential, air change rates and air cleanliness, both viable and particulate levels).

Although repairs and retrofitting are possible to upgrade these systems, serious consideration should be given to how much basic capacity exists in the system and the construction quality of the components before you add, for example, extensive numbers of aseptic manufacturing areas to a plant which was originally designed for biocontainment only. Completely new construction or the use of modular cleanrooms might be a more efficient use of resources, especially in terms of validation.

7..3 PHARMACEUTICAL WATER SYSTEMS

- Literally, the "main ingredient" in most biologics processes/products
- A "no compromise" issue with regulatory authorities
- High maintenance system
- Requires extensive monitoring and documentation

There are three recognized water qualities in the United States, Potable Water, Purified Water and Water For Injection (WFI). Potable Water is basically drinking water, and standards are prescribed by public health and environmental authorities. Purified water and Water For Injection (WFI) standards are defined by USP monographs. Purified water is usually produced by deionizing systems, such as reverse osmosis. It is primarily used for non-parenteral products and in processes well before the production of a final sterile dosage form, such as fermentation [13].

Water For Injection is in a class all by itself. Because of its extreme purity, WFI is an extremely aggressive solvent. Normally, it is kept at 80 degrees C in a continuously recirculating loop. As it can be used in parenteral products, freedom from microbial contamination essential. *Under no circumstances should a contaminant be added to the system as part of a validation program.* For retrospective validation, system design, documented construction information and IQ, OQ protocols would have to be available. Due to the complexities and integrated nature of a WFI system, a piecemeal program of upgrades will not result in a "validatable" system. Any claims of "upgrading" should be viewed skeptically.

7.4 MULTI USE FACILITIES

Unless a facility is intended for a specific product, manufacturers should plan to operate and validate as a *multiuse* or *multiproduct* facility. Major considerations for this type of operation can be summed up as "4-C's".

- Campaigning
- Cleaning
- Clearance
- Closed systems

Unless a multiuse facility is designed and constructed to have a number of completely separate venues, a system of *campaign*ing is the best solution to minimize the possibilities of product cross contamination or mix-up. Campaigning is considered a sequential use of a facility or equipment, with only one product in the facility at a time. Such a procedure calls for excellent cleaning and quality assurance programs. Although this may seem like a novel idea, all GMP operations have always required clear demarcations even between production lots of the same product.

A manufacturer must validate *cleaning* procedures for each piece of common use equipment or production line to ensure that any residual product from the previous run had been removed to be below detectability standards set by the manufacturer and accepted by Regulatory Authorities. There is precedent for manufacturers to validate a generic process for many products. Examples are dry heat depyrogenation or the use of caustic and acid to clean chromatography columns. Manufacturing managers should ensure that validated cleaning procedures have been determined before new products are introduced into production.

Clearance must be approved or conduced by the Quality Assurance Unit. A detailed protocol, establishing limits must be approved and in place before any area clearance can be given. Written evidence of area clearance should be included in every batch record.

Finally, *closed systems* and disposable products minimize the magnitude of facility and equipment cleaning and validation. However, if a system is to be considered "closed" it must be validated to prevent escape of product as well as entry of contaminants during operation.

8. Glossary

Change Control - A formalized program by which qualified representatives review proposed and actual changes to products, processes, equipment or software to determine their potential impact on the validation status.

Acceptance Criteria - A set of measurable qualities or specifications used to check out system or equipment installation, operation or performance. Conformance with these qualities or specifications provides a high degree of confidence that a system is installed, operates or performs as intended.

Concurrent Validation - Establishing documented evidence that a process does what it purports to do based on information generated during actual implementation of a process.

Installation Qualification (IQ) - Documented verification that all key aspects of the installation adhere to appropriate codes, approved design specifications and manufacturer's recommendations (where appropriate).

Operational Qualification (OQ) - Documented verification that systems or subsystems, capable of being challenged, perform as intended throughout all anticipated operating ranges.

Performance Qualification (PQ) verifies the performance of critical utility systems or processes. Critical utility systems such as WFI and Pure Steam are challenged throughout proposed operating ranges for extended periods of time, while an extensive program of quality monitoring is performed.

Process Validation - A defined strategy of inter-related practices and procedures which in combination with routine production methods and quality control techniques provides documented assurance that a system is performing as intended and/or that a product conforms to its pre-determined specifications.

Prospective Validation - Establishing documented evidence, prior to process implementation, that a system does what it purports to do based on a preplanned protocol.

Retrospective Validation - Establishing documented evidence that a system does what it purports to do based on review and analysis of historical information.

Turnover Packages (TOP) - Documentation binders or files which contain all relative design, installation and testing documentation generated during design and construction for a particular engineered system. This document package is intended to support the Installation Qualification. (May be prepared by contractor)

Validation Master Plan - A technical and administrative resource planning tool for the validation and commissioning process and to provide a guideline for the validation approach for each piece of equipment and process.

9. Sample Facility Validation Master Plan

(This section was adapted from an unpublished validation master plan for Building 459 at Fort Detrick [15])

9.1 INTRODUCTION

9.1.1 *Purpose*

The primary purpose of this Validation Master Plan is to provide a technical and administrative resource planning tool for the validation and commissioning process and to provide a guideline for the validation approach for each piece of equipment and process. The document also provides facility and organizational descriptions which will be used to familiarize regulatory agencies, employees and consultants with the facility and its validation program.

This document is not a validation protocol. The test functions and acceptance criteria given within this document are proposed and intended to provide a starting point and a checklist for preparing specific facility, equipment, process and procedure validation protocols. Test functions, test instructions and acceptance criteria will be formally approved by approval of specific Installation, Operational and Performance Qualification protocols. The plan may be revised as necessary during the design, commissioning, and validation phases of the project, but will not be revised after the initial facility validation.

9.1.2 *Scope*

This Validation Master Plan applies to the [Name] Facility located in building [Number] at [Location] It does not apply to other laboratories or facilities operated by [Firm].

9.1.3 *Document Organization*

In order to evaluate the validation approach, an understanding of the manufacturing facility is required; consequently, Sections 9.2 and 9.3 of the plan describe the manufacturing processes, equipment and facilities. These sections also include air, personnel and material flow descriptions.

Section 9.4 describes the proposed Quality Assurance Plan for products manufactured in the new facility. This includes the organization of the Quality Assurance and Quality Control Departments and their relationship to manufacturing.

Section 9.5 describes the validation management and administrative organization for the start-up and commissioning of the facility.

Section 9.6 identifies the overall approach and documentation to be used and the required validation protocols for the facility and equipment. It also includes summaries of the proposed test functions and acceptance criteria and/or the specific approach for validation of processes. It does not include the required protocols for validation or verification of analytical methods; however, the general strategy for validation is described.

9.2 PROCESS DESCRIPTIONS (To be developed)

9.3 FACILITY AND EQUIPMENT DESCRIPTION

9.3.1 *Unit Operation Area (Examples)*
- Receiving, Inspection, Quarantine and Release Area
- Component and Glassware Preparation
- Media Preparation Area
- Cell Culture Area
- Cold Rooms
- Purification Suites
- Aseptic Processing Staging
- Filling / Vialing Suite
- Finished Product Packaging, Quarantine and Release Areas
- Quality Control and Stability Area
- Containment Laboratory
- Biochemistry Research Laboratory
- Waste Decontamination Area

9.3.2 *Material and Personnel Flow (Examples)*
- Product Flow
- Personnel Flow
- Glassware Flow
- Waste Flow

9.3.3 *HVAC Zone Identifications and Air Flow and Pressurization*

9.3.4 *Utility Systems (Examples)*
- Water For Injection System
- Pure Steam System
- Compressed Gas Systems
- HVAC Systems
- Heat Exchange Systems
- Controlled Temperature Environments
- Waste Decontamination

9.3.5 *Process Equipment (Examples)*
- Liquid Nitrogen Freezers
- Incubators/Freezers/Refrigerators
- Biosafety Cabinets
- Media Preparation Equipment
- Cell Culture Fluid Harvesting Equipment
- Purification Equipment
- Sterile Filtration Systems
- Depyrogenation Oven
- Autoclaves
- Glass, Vial and Stopper Washers

- Filling, Stoppering, and Capping Equipment

9.3.6 *Automated Equipment and Computer Related Systems*

9.4 QUALITY ASSURANCE PLAN

9.5 VALIDATION PROJECT MANAGEMENT

9.5.1 *Organizational Chart*

9.5.2 *Validation Responsibilities*

[Firm] has overall responsibility for designing, constructing, commissioning, validating, operating and maintaining the facility as well as manufacturing, quality assurance, quality control and release of the products. [Firm] has hired an architecture/engineering company to design the renovations for the facility. [Firm] has also hired a design and validation contractor to provide design review and validation services. [Firm] will hire a construction company to manage the construction.

The validation related responsibilities of the various parties involved in the project are also identified in Figure 9.5.1.

9.5.3 *Validation Staffing*

The project team which will prepare protocols, execute protocols and prepare reports will be a combination of Staff members and [Validation Firm]'s validation specialists. In order to assure that all work is performed and approved by qualified individuals, the Responsible Head of the Facility (or Firm), will assign responsibility for validation of specific systems or items and/or responsibility for review and supervision of validation work to members of the project team based their experience, education and training.

The Responsible Head of the Facility and the Quality Assurance Manager will have responsibility for approval of all validation protocols, final reports, and standard operating procedures as well as other critical GMP documentation. The Responsible Head of the Facility will also have the responsibility for determining which additional staff or consultants must review and/or approve specific protocols or final reports.

9.5.4 *Project Schedule*

9.5.5 *Document Control (Examples)*

- System/equipment Turnover Packages
- Calibration Certificates
- Laboratory Test Records
- Validation Protocols and Final
- Standard Operating Procedures
- Document Numbering Systems
- Equipment Numbering Systems

9.5.6 *Design and Validatability Review*
The validation of the facility started with the simultaneously with the initial conceptual design. At key stages throughout the design process, design review meetings were held to review and make key GMP design decisions. The meetings were attended by the Responsible Head of the Facility, architects and engineers from the [Architecture and Engineering Firm], architects, engineers, compliance specialists and validation specialists from [Validation Firm], scientists, Quality Assurance Management, maintenance and design engineers, Safety and Environmental Specialists and independent GMP consultants.

Throughout these design/review meetings, GMP and functionality requirements have and continued to be determined for each aspect of the facility including, but not limited to, the following areas:

- Architectural layout, Personnel and Material Flow
- Architectural Finishes / Room Criteria
- Purified Water and Steam
- HVAC and Clean Room
- Clean Compressed Gasses
- Mechanical Equipment
- Biological Containment
- Material Receipt, Storage and Shipping
- Quality Assurance/Control Testing and Inspection
- Drainage System
- Process Equipment Design
- Maintenance and Calibration
- Emergency Power
- Safety Engineering
- Waste Decontamination
-

After the requirements for each aspect are determined, the responsible party (either the Architectural and Engineering Firm, or the Facility) completes design and it is again reviewed by the project team and the design review consultant [Validation Firm]. After the approved designs are included in the construction specifications, the validation life cycle continues with the preparation and execution of the validation documents.

9.5.7 *Validation Documents*
Turn-Over Packages. Turnover Packages (TOP) are documentation binders or files which contain all relative design, installation and testing documentation generated during design and construction for a particular engineered system such as HVAC or Pure Steam. TOPs are analogous to Owner's Manual or Operators Manual which would be provided with stand alone package equipment such as stopper washers or freezers. TOPs serve to capture all engineering documentation generated throughout

the design construction process. As such, their completion can be included in construction or purchase bid specifications. This information is compiled into single source packages which then provide the foundation for Installation and Operational Qualification.

The TOP for each system will contain all information relating to the following areas:

- General / Introduction / Scope
- Equipment
- Piping
- Instrumentation
- Controls and Automation
- Electrical
- Operating Procedures
- Maintenance Procedures and Schedules
- Calibration Procedures and Schedules
- Calibration Specifications (Accuracy Requirements)

The Turnover Packages will be prepared for each of the following systems:

- WFI System
- Pure Steam System
- Drainage and Waste Decontamination Systems
- Compressed Gas System
- Electrical Systems
- Building Automation System
- HVAC System

Installation Qualification Protocols. The Installation Qualification (IQ) verifies that the equipment or system and its corresponding utilities are installed in accordance with design specifications, manufacture's recommendation and GMPs. In addition the IQ will confirm that critical instruments are calibrated, and that system components are properly identified. Any exceptions will be documented, corrected and/or justified.

Where possible and applicable, pre-shipment inspection of equipment and documentation is included as part of the installation qualification. Pre-shipment inspection is performed along with vendor audits to address issues such as software development and quality assurance plans, operational reports and specific vendor/purchaser inspection reports.

Typical installation qualification test function will include, but are not limited to the following:

- Purchase Order Specification
- Manufacturer Specifications

- Piping and Instrument Drawing
- Construction and Installation
- Instrumentation Identification and Calibration
- Microprocessor Configuration
- Test Equipment Calibration
- Required Spare Parts
- Utility Systems Capability
- Product/Water Contact Materials of Construction
- Passivation and Cleaning
- Weld Inspection
- Line Slope
- System Installation Compliance to GMPs

The Installation Qualification is performed in accordance with a pre-approved written protocol. Specific Installation Qualification attributes to be verified during the performance of the Installation Qualification protocol will be developed from the system and component design criteria, GMP requirements and other specifying documentation. They, along with the acceptance criteria, will be approved as part of the protocol approval process.

Operational Qualification Protocols. The Operational Qualification (OQ) verifies that the system or equipment operates in accordance with the design specifications, manufacturer's recommendations, and GMPs. OQ testing is designed to simulate the full range of operating conditions to establish a system capabilities baseline and to assure the system operates as intended.

Operational qualification test functions include but are not limited to the following:

- General Operational, Control and Interlocks Verification
- Functional Testing of Computer Related Systems
- Specific Operational Testing (generation of baseline data)
- Utility System Capability
- Calibration of Test Equipment
- Standard Operating Procedures
- Maintenance Procedures
- Cleaning or Sanitizing Procedures

For equipment or systems controlled or monitored by a computer related system, the functional testing is included and performed as part of the operational qualification. Functional testing verifies that the integrated hardware and software program perform in accordance with the functional specifications developed during the Requirements Phase.

The Operational Qualification is performed in accordance with a pre-approved written protocol. The specific Operational Qualification attributes to be verified during the performance of the protocol will be developed from the design specifications, manufacturer's documentation, GMP requirements and other specific documentation. They, along with the acceptance criteria, will be approved as part of the protocol approval process.

Change Control Initiation. Each system\equipment\process will be covered by the Engineering Change Control program (See section 9.7.1) immediately following the completion of the Operation Qualification test functions. Any changes made to the system during Operational Qualification must be documented in the Final Report. Any engineering changes proposed after the Operational Qualification (i.e. during the cycle development, performance qualification or manufacturing) must be pre-approved and follow the Engineering Change Control Procedure. If changes are made prior to approval of the Final Report, the change control documentation must be included in the Validation Packages and the change and justification for change must be document in the Final Report.

Cycle Development. Critical processes other than the actual product manufacturing process such as sterilization, component washing, and equipment cleaning require pre-validation cycle development work. This may include establishment of appropriate sterilization cycle types, spray ball configurations, cleaning agents etc. This test work is performed during or after operational qualification. Cycle development does not have to be performed per a written procedure; however, the procedure and results of all work must be recorded in controlled research laboratory notebooks, performed in accordance with GMP documentation practices and sound scientific methods, and retained as part of the Validation Package. At the completion of cycle development work, a brief cycle development report will be prepared which identifies parameters that have a high probability of successfully meeting the performance qualification requirements

Performance Qualification Protocols. The Performance Qualification (PQ) verifies the performance of critical utility systems or processes. Critical utility systems such as WFI and Pure Steam are challenged throughout proposed operating ranges for extended periods of time, while an extensive program of quality monitoring is performed. Critical processes such as sterilization of components are challenged three (3) or more consecutive times under worst case conditions.

The Performance Qualification is performed in accordance with a pre-approved written protocol. The specific Performance Qualification attributes to be verified during the performance of the protocol will be developed from the finished product specifications, research and development data, GMP requirements and other specific documentation. They, along with the acceptance criteria, will be approved as part of the protocol approval process.

Process Validation Protocol. The Process Validation (PV) verifies the performance of the overall product manufacturing process. Process validation is performed on the entire product manufacturing process which includes all support processes, preparation of media, components, buffers, formulation, filtration, filling and packaging. During process validation, three consecutive batches are manufactured and all in-process and finished product specification are verified and much higher than normal inspection quantities and frequencies. Critical parameters are set at different ranges during the manufacturing process to verify that the product consistently meets its pre-determined quality specification over the range of critical parameter settings. Consequently, a proven acceptable range for the each critical parameter is developed.

All equipment, processes, quality assurance test procedures must be validated (or verified if compendial method) prior to the start of Process Validation.

The Process Validation is performed in accordance with a pre-approved written protocol. The specific Process Validation attributes and process parameters to be verified during the performance of the protocol will be developed from the finished product specifications, research and development data, the manufacturing process capability, GMP requirements and other specific documentation. They, along with the acceptance criteria, will be approved as part of the protocol approval process.

Validation Final Reports. The Final Report summarizes the results of the validation process and provides an analysis of the test data in support of the conclusion that the equipment or system demonstrates consistent performance within the established acceptance criteria. Deviations and/or exceptions to approved protocols, along with suitable explanations, are also documented in the final report.

Validation Package. The final report, completed protocols and all supporting documentation is then gathered into a Final Report Package for review, approval and retention.

Certification for Use in Manufacturing. Validation of a system/equipment/process will be considered complete when its respective protocols identified in section 5 have been successfully implemented and the Final Report has been approved. A system\equipment\process may be certified for use in manufacturing when the following have been completed:

- Applicable Protocols have been successfully completed and the Final Report has been approved.
- Applicable SOPs have been approved.
- Validation is complete for all utility systems which support the equipment or process and for all equipment that supports a process.
- All system\equipment\process instrumentation is still within calibration.

9.6 REQUIRED DOCUMENTATION

9.6.1 *Required Protocols and Procedures*

A matrix will be developed to include each system, piece of equipment and process to be validated. The purpose of the matrix is to serve as a management and organizational tool. The amount of detail and specificity in this matrix is a decision for the authors. Examples of recommended documentation for inclusion in the matrix are listed below.

- IQ - Installation Qualification Protocol
- OQ - Operational Qualification Protocol
- CD - Cycle Development Study and Report
- PQ - Performance Qualification Protocol
- PV - Process Validation Protocol
- OP - Operating Procedure
- PM - Preventive Maintenance Procedure
- CL - Cleaning Procedure
- CAL - Calibration Procedure

9.6.2 *Validation of Utility Systems*

This section would provide a summary of the key validation test functions and acceptance criteria for each utility system.
Refer to section 9.6.1 for the required protocols for each system.

9.6.3 *Qualification of Process Equipment*

This section would provide a summary of the key validation test functions and acceptance criteria for process equipment.
Refer to section 9.6.1 for the required protocols for each system

9.6.4 *Validation of Support Processes*

This section would provide a summary of the key validation test functions and acceptance criteria for each support process.

9.6.5 *Aseptic Filling Validation*

Protocols will be developed to demonstrate that the product is aseptically filled into a final dosage container.

9.6.6 *Cleaning Validation*

Cleaning validation protocols will be developed for each piece of equipment used for each product to be manufactured.

9.6.7 *Purification Validation*

The purification process validation will be comprised of the qualification of several interrelated processes and the chromatography system equipment. The purification process for each unique product will be validated; however, processes within the purification scheme for a new product which have previously validated will not be repeated.

9.6.8 *Computer Related System Validation*

Computer related systems used in manufacturing and quality control for the Facility will be validated using the approach described in several relevant industry documents including the Federal Information Processing Standards Publication 101 (June 6, 1983), PMA Computer Validation Committee's Validation Concepts for Computer Systems Used in the Manufacture of Drug Products, the Parenteral Drug Association's Draft Report on Validation of Computer-Related Systems and articles and speeches prepared by FDA investigators.

Computer System Validation places a large emphasis on the software development process in an effort to build in a quality. Two different computer validation situations will be encountered in the Facility. The first situation involves packaged equipment such as autoclaves or chemical analyzers which are controlled or monitored by computer related systems. Different equipment may be supplied with various amounts of validation testing. Some equipment will be supplied fully validated by the manufacturer while other equipment may have software which has been developed in a validatable manner but requires some degree of verification and testing. The validation documentation provided by the equipment manufacturer will be audited for compliance to Computer System Validation and the number and detail of the test functions and acceptance criteria required for each piece of equipment will be based on the degree to which previous validation procedures had been completed.

The second situation is the Building Monitoring System which will be designed by the architecture/engineering firm and developed by a specialty vendor. Because this is a customized system specifically designed for a particular system or facility, all phases of Computer Validation can and will be performed concurrently with the design and development activities.

9.6.9 *Quality Assurance/Control Laboratory Validation*

- Laboratory Equipment Qualification
- Computer Related Systems used in the QA/QC Laboratory
- Compendial Method Verification
- Novel Method Validation

9.6.10 *Process Validation*

The process validation for the overall manufacturing process will be performed as indicated in Section 9.2 for each new product. In-process and finished product sampling will be frequent and adequate to assure the lot does not contain non-conforming units. Each manufacturing lot will be produced in accordance with a batch record as well as a validation protocol. In addition, research and development data will be used in conjunction with laboratory scale test work to support validation of each process.

9.7 OTHER GMP PROCEDURES AND PROGRAMS

9.7.1 *Engineering Change Control*

- Equipment

- Software Revision Control

9.7.2 *Revalidation*

9.7.3 *Calibration*

- Standards
- Instrument Numbering System
- Instrument Calibration Files

9.7.4 *Preventive Maintenance Program*

9.7.5 *Standard Operating Procedures*

9.7.6 *Facility Cleaning and Sanitization*

9.7.7 *Environmental Monitoring Program*

9.7.8 *Training*
A GMP training system will be developed which ensures that all personnel are trained in current Good Manufacturing Practices as required by Title 21 CFR Part 211.25(a) and in all operations which they are required to perform. The program will be designed so that training is performed by qualified personnel. All training will be documented and retained in a training file.

9.7.9 *Batch Production Records*

9.7.10 *Equipment Log Books, Status Tags and Room Clearance Checklists*
A system for determining the day-to-day status of critical systems, areas and activities will be developed. Activities such as batch production, cleaning, preventive maintenance, calibration, load sterilization, etc., will be controlled and documented in logbooks by procedure and reviewed and filed as critical processing documentation.

Status tags will be utilized to indicate the status; such as dirty, in-process, clean, quarantine, product manufactured, etc. Procedures will be developed for the placement and removal of status tags.

Checklists will be developed to establish proper room clearance prior to initiating a production batch. Checklist development, handling, control and review will be incorporated into procedures for on-going control.

9.7.11 *Equipment History Files*
A program to ensure control of critical system documentation will be established. The files will be controlled by Quality Assurance or other appropriate department and centralized for easy retrieval. Data such as operating and service manuals, purchase orders, manufacturers' specifications, as built drawings and schematics, spare parts

lists and any other information pertaining to the system will be included in the file. Procedures will be developed to create files for new equipment, update information for existing equipment, control, removal and return of information, etc.

9.7.12 *Label Control Program*

9.7.13 *Facility Pest Control Program*

9.7.14 *HEPA Filter Integrity Testing*
A procedure will be developed which identifies the certification requirements for all High Efficiency Particulate Air (HEPA) Filters in the Facility.

9.7.15 *Filter Integrity Testing*
A filter testing procedure will be developed to ensure bacterial retentive air and liquid filters are integrity tested and replaced as necessary. This program will cover those filters not directly in the process stream such as vent filters on tanks and autoclaves, and compressed gas filters. This procedure will include filter integrity testing instructions, schedules, and acceptable test results.

Filters directly in the process stream such as product sterilizing filters, media sterilizing filters and harvesting ultrafilter will be tested before and after the processing of each batch and documented in the actual batch production records.

9.7.16 *Supplier Auditing and Qualification*

9.7.17 *Security and Access Control*

9.8. GLOSSARY

9.9 DRAWINGS

9.9.1 *General Floor Plan*

9.9.2 *HVAC Air Handling Unit Zone Identifications*

9.9.3 *HVAC Air Flow Diagrams*

9.9.4 *Personnel Flow*

9.9.5 *Glassware Flow*

9.9.6 *Waste Flow*

9.9.7 *Individual Product Flow Diagrams*

PHARMACEUTICAL AND REGULATORY ASPECTS OF CONVERSION

J. MELLING
The Salk Institute for Biological Studies
Biologicals Development Center
PO Box 250 Swiftwater PA 18370
USA

1. Introduction

When considering the issue of conversion of former BTW facilities there are a range of possibilities and account must be taken of the nature of the facility to be converted as well as the desired outcome of the conversion process. The process of conversion is likely to be funded by international or national agencies and because of the political and international security benefits this activity will not be subject to normal market forces. However, if the end result of conversion is a facility which produces products or services for sale then eventually such an entity will be exposed to the competitive market and will survive or fail according to its economic viability. Thus before committing to any particular conversion activity the planned outcome should be carefully evaluated in order to ensure that it is both technically feasible and that the new entity to be created will indeed be viable.

Having the creation of a pharmaceutical facility as the goal of conversion has at first sight significant attractions not the least of which is the humanitarian ideal of converting life threatening BTW facilities into ones which produce lifesaving medicinal products.

Moreover because the underpinning technologies such as microbiological techniques, fermentation and downstream processing are superficially similar, it is tempting to assume that buildings, equipment and people can as easily serve the new as they did the old. It needs to be recognized however that attempting to change a BTW facility into one which can produce pharmaceutical products for human or animal use will be highly challenging because of the exceedingly rigorous standards of operation which medicinal product manufacture requires. There will also be issues affecting long-term survival as to whether or not such a new player in the pharmaceuticals market place can carve out an economically viable niche and create a sustainable product portfolio.

A final point is that the whole aim of conversion is to move so far from the original offensive purpose of a BTW facility that there can be no going back. Thus the retention

E. Geissler et al. (eds.), Conversion of Former BTW Facilities, 107-117.
© 1998 *Kluwer Academic Publishers. Printed in the Netherlands.*

of capabilities to work with pathogens on any significant scale will always remain a concern.

2. Candidate Facilities for Conversion

The term BTW facility may include all of the elements which together make up a capability to produce biological weapons. The most obvious are Research, Development, Production, Storage, Weaponization, and Testing. While some elements would, by their very existence, contravene the convention others would not be prohibited per se providing that their end product was not for offensive use .

Because of the gray area which can exist between legitimate vaccine, therapeutic and diagnostic uses of microbiological facilities and prohibited activities it will be advisable to limit the microbial agents being worked on so as to exclude all those considered to have an offensive potential. It is clearly acceptable for any country to retain a capability to produce defense vaccines and reagents for detection and diagnosis of BTW agents for defense. It would be preferable however that such activities are kept separate from normal medicinal product production so as to increase transparency especially in the conversion process.

Table 1 sets out the "most likely" BTW agents as exemplified in the US Department of Defense list of required vaccines.

TABLE 1. BW agents and diseases of concern

Anthrax, Botulinum toxins,Q-Fever, Ricin toxin, Smallpox, Staphylococcal enterotoxin, Venezuelan Equine Encephalomyelitis, Western Equine Encephalitis, Eastern Equine Encephalitis, Brucellosis, Plague Tularemia

Data derived from the US Department of the Army, Joint Program Office for Biological Defense Request for Proposal DAMD17-95-R-5020 for manufacture of defense vaccines

None of the agents listed now causes such significant levels of disease in the general population as to require mass population immunization and accordingly any work involving such agents in a converted facility would have to be well justified and transparently non-offensive.

2.1 RESEARCH FACILITIES

Clearly research facilities do not in themselves constitute a BTW capability. However in a conversion program aspects of the facility such as the level of containment, seed stocks of microorganisms, types of equipment, operating procedures and work programs must be demonstrably those required to support a pharmaceutical as opposed to a BTW program.

2.2 DEVELOPMENT FACILITIES

The scale of operations in development facilities can range from small laboratory equipment to levels approaching those which, depending on the microbe involved, could produce significant quantities of hazardous material. Thus there should be a requirement to limit and justify the types of equipment, biocontainment levels and the actual work undertaken in terms of the pharmaceutical program objectives. Because of the likely previous use of development facilities an essential prerequisite of conversion to pharmaceutical operations would be decontamination, clean-up and most importantly validation that such operations had been effective. This is discussed in detail in section 3 below.

2.3 PRODUCTION FACILITIES

These are likely to contain large-scale fermentation and down-stream processing equipment and it is probable that the operating scale may not be appropriate for the new purpose. Retention of over-capacity should be avoided as should retention of any biocontainment capability which is not consistent with the new purpose. Again, ensuring that such a facility can be made to comply with pharmaceutical production requirements will be critical to determining whether or not conversion is feasible (see section 3).

2.4 STORAGE FACILITIES

Such facilities would have been used, or designed to be used for agent stockpiling prior to weaponization. It is likely that these would have the capability for cold storage and probably be underground and/or in hardened bunkers.

Such facilities would not be appropriate for conversion to a pharmaceuticals production program and should be destroyed.

2.5 WEAPONIZATION FACILITIES

Clearly these could contribute nothing to a converted operation and should therefore be dismantled and destroyed.

2.6 TESTING FACILITIES

These are likely to be of two types; either laboratory or field facilities.

2.6.1 *Laboratory Testing Facilities*

Laboratory facilities could include general animal laboratory facilities including biocontainment animal holding and in addition aerosol generating equipment and exposure chambers. Animal test facilities form a key part of a pharmaceutical production operation and the issue would be how appropriate the biocontainment capability might be. Equipment for aerosol generation and exposure is not required for the majority of medicinal product development or manufacture. Thus retention of such a capability would require stringent justification. Once again the difficulty arises that the more a pharmaceutical facility engages in work involving products, including vaccines, derived from infectious agents or toxins the less convincing the conversion will be to the world at large.

2.6.2 *Field Testing Facilities*

These have no relevance to pharmaceutical operations and should therefore be dismantled with appropriate surveillance to ensure that no reversion to BTW activity takes place.

3. Constraints on Using Converted BTW Facilities for Pharmaceutical Production

Producing pharmaceutical products is, rightly, one of the world's most heavily regulated industries with operating regimes being enforced by periodic inspection by government agencies. Until the 1960's the absence of such strict controls produced a number of tragic deaths and disabilities as well as probably allowing the release of ineffective although "safe" products. Converted BTW facilities must therefore be designed to operate in full compliance with those regulations covering the types of products they intend to produce.

It is likely that the pharmaceutical products being made in converted BTW facilities will be biologicals. Biological [1] products are defined by the following criteria:

1). They may either have no clearly chemically defined composition or because of their tertiary or quaternary structure, simple chemical analysis is insufficient for effective characterization.

2). The only way of qualitatively or quantitatively evaluating them usually is by measuring their effect in vivo.

3). Quality cannot be guaranteed from a final test on random samples, but only from a combination of in-process and end-product tests that monitor the entire manufacturing process.

One consequence of this characteristic is that in vivo assays are required. Another and more far reaching one is that because of the interrelationship between the method of production and the characteristics of the final product, stringent control on all aspects of the production process is required in order to ensure reproducibility of the end-product. This contrasts with many chemical drugs where methods of manufacture can vary to a fairly wide degree so long as the end-product meets certain specifications. For biological products on the other hand, the same degree of certainty of quality can only be achieved by a much stricter control of the production process. For biologicals the product is indeed defined by the production process.

Accordingly, a wide range of standard operating procedures (SOPs) must be created to address every step of the production process including support service and maintenance operations. Such disciplines which encompass facility planning, equipment design and process operations will be foreign to BTW facilities and must be taken into account in planning any conversion.

If facilities are intended to produce more than one product, i.e., multiuse, there then will be additional operating requirements to avoid product cross-contamination. For economic reasons multi-use is likely to be a desired outcome.

3.1 BUILDINGS AND ASSOCIATED SERVICES

This is not intended to be a detailed account of how to design buildings and services for pharmaceutical bioprocessing, but rather an attempt to illustrate the type of problems which are likely to be encountered in a conversion process.

The essential principle in the design of a facility for production of biological products is to minimize the potential for mix up of one product with another or contamination of finished or late stage product with starting materials. Thus segregation, both physical and temporal, of different products and stages of manufacture is vital. Physical segregation is achieved by designing the facility so that product flow patterns ensure there is no cross over between early and later stages of processing. In addition the building design should address the movement of personnel, raw materials, equipment and waste materials. Normally these flow control concepts are part of the building design and retroactive adaptation of existing buildings is likely to present problems. Isolation of different manufacturing stages is further aided by having air handling equipment which is dedicated to specific manufacturing areas such as fermentation, down stream processing, filling and lyophilization. Air pressure differentials between the various processing areas also contribute to effective segregation. Clearly the quality of the air supplied to process areas is critical in minimizing opportunities for either microbial or particulate contamination.

Cleaning is of major importance and surfaces of walls and floors should be non-porous so as to be resistant to frequent exposure to materials used for cleaning and/or decontamination. To aid cleaning, surfaces should not have ledges or cracks and pipework or other services which are hard to clean should be enclosed to avoid a build up of particulates.

There are stringent criteria for the quality of steam and water used in bioprocessing which may go into the product or have contact with equipment used in the production process. There is therefore a requirement for pure steam generators that are designed for the production of Water for Injection (WFI) quality pyrogen free steam for use in pharmaceutical and biotechnological processes.

Plant steam is unacceptable for use in the previously mentioned applications. This is due to the presence of contaminants including boiler treatment additives along with the potential for pyrogenic contamination. A pure steam generator can use the plant steam as a primary energy source because the plant steam never contacts the pure steam.

Once WFI has been generated its quality, especially freedom from microbial contamination, must be maintained and normally this is done by installing a continuously circulating loop maintained at 80°C. The system design and operational validation are essential to obtain regulatory approval.

Air quality appropriate to the particular process area is achieved by HEPA filtration of incoming air and also ensuring by the number of air changes and pressure differentials, that the particulate count is maintained below the relevant limits.

To meet all of the necessary standards by retrofitting an existing facility poses significant problems and careful design and planning are essential.

3.2 EQUIPMENT

The issues relating to equipment are essentially the same as for buildings in that the aim is to prevent cross contamination or confusion of one product with another. In the case of measuring equipment to ensure that the readings are correct, such equipment must be validated at appropriate intervals.

3.2.1 *Shared Equipment*

This is equipment which is used with different products and/or at different process stages with the same product. The essential requirement is to have in place validated cleaning or decontamination procedures and to have SOPs for the changeover operations. Typically items such as fermentors, centrifuges, lyophilizers and filling machines fall into the category of shared equipment.

Equipment (or buildings) which have been used for work with spore forming organisms cannot subsequently be used for pharmaceutical production of products from non-sporeformers[2].

A major difficulty to be overcome with any equipment which previously had a BTW use will be clearly to demonstrate that microorganisms or toxins to which it was exposed have been inactivated and removed. In view of the likely agents involved there will be great concern lest any pharmaceutical products become contaminated. If clean-up and decontamination cannot be validated such equipment will be of no use.

3.2.2 *Dedicated Equipment*

This may only be used for one process stage of one product although it can be re-used for the same process stage of subsequent batches of the same product. Obviously cleaning is required, but the requirements are less onerous than was the case with shared equipment. By definition no equipment previously used in BTW operations will be acceptable.

3.2.3 *Disposable Equipment*

Such equipment is discarded after use and mainly consists of pipettes, syringes and various plastic or glass items.

3.3 PERSONNEL

It cannot be said too often that pharmaceutical manufacture is a highly disciplined and heavily documented activity. It is essential, as past problems have demonstrated, to minimize the possibility of human error by using SOPs and check sheets to ensure compliance. There is thus a requirement for personnel to be trained over and above their technical capabilities in order to ensure that they can work effectively and comply with all of the aspects of current Good Manufacturing Practice (cGMP)[3,4]. Training is not a once and for all activity, but needs to be repeated at intervals and the training status of all employees recorded.

Experience has shown that attempts to convert research scientists to work in accordance with cGMP is very difficult. There is probably a negative correlation between creative research ability and the capability to function well in a cGMP environment. Thus an important part of a successful conversion process will be personnel selection, training and development. It is likely that there will have to be new hiring to bring in key specialists with cGMP experience in order to make a success of conversion.

3.4 OPERATING PROCEDURES

In a pharmaceutical facility every activity which can impact on product quality needs to be subject to an SOP. As has been referred to above this requirement goes well beyond the production process itself and also covers most aspects of the facility's operation .

Without direct experience of operating procedures in a BTW facility it is difficult to be certain, but it is most unlikely that the operating disciplines would approach those required for cGMP. In the process of conversion therefore a mass of documentation will have to be created de novo to provide the necessary process regulation. Even in countries where cGMP is the norm it still proves difficult for new entrants to the pharmaceuticals business to achieve the necessary standards. Accordingly the steepness of the learning curve involved in moving from BTW to pharmaceutical activities is likely to be daunting.

4. Product Licensing Issues

There are three groups of biological products which a pharmaceutical bioprocessing facility may generate; vaccines, therapeutics and diagnostics. The first two are subject to intense regulation, but even diagnostics are controlled since a mis-diagnosis can be as fatal as an ineffective product.

Thus in addition to the facility complying with appropriate standards, the product(s) which are generated will be subject to intense regulatory scrutiny.

The criteria which such products must meet involve: (a) safety tests, including toxicological assessment, initially in animals, but then in man; (b) determination of efficacy, again initially using animal models, followed by human studies, and (c) assessment of product quality, involving data on all aspects of production designed to ensure not only that the product is acceptable according to current scientific standards, but also that the standard of quality can be maintained and reproduced from batch to batch during the life of the product.

4.1 SAFETY

A number of elements are involved in ensuring the safety of biological products. Seed lots of the particular organism must be established and subjected to careful scrutiny as to the identity and provenance of the organism. This is especially important for live attentuated vaccines, but also for other biologicals to ensure the presence of appropriate macromolecules.

The various ingredients used in growth media and materials with which the product comes into contact during processing must be selected with a view to their potential toxicity. This is especially so where carry-over of minute and undetectable amounts into the final product could have adverse consequences. For example, since the recognition of bovine spongiform encephalopathy (BSE), materials of bovine origin, particularly from BSE endemic areas, are excluded. This has required reformulation of many culture media for both bacteria and viruses, as well as reappraisal of seed stocks which may have been produced using bovine products. A further example is the need to eliminate the use of toxic chemicals at the processing stage and to set limits for carry-over of other materials and determine that such limits are well below any toxic levels. Materials originating from living organisms which may be used in the production or purification processes (e.g., animal sera and monoclonal antibodies) must either be demonstrably free from adventitious microorganisms or it must be shown that the procedures used would eliminate the risk of such organisms being carried through into the final product.

Depending upon the type of biological product, particularly vaccines, appropriate tests including animal studies are required to validate the inactivation process or ensure effective attentuation. In the case of inactivation of organisms or toxoiding of toxins, the kinetics of inactivation must be determined to ensure a sufficient margin of safety is built into the process. Because any change in formulation or conditions (e.g., pH, temperature, ionic strength) of inactivation can radically alter the kinetics, great care is needed to ensure reproducibility.

Once a production process has been established and animal safety data established, tests in man would be the next stage.

Phase I studies in healthy adult volunteers would be required to show absence of unacceptable side-effects before proceeding to Phase II studies, again with informed consent, in larger numbers of people in the target population for the product. Normally only after acceptable Phase II studies for safety would larger Phase III trials for efficacy be undertaken; these studies would also contribute to the data on adverse reactions.

4.2 EFFICACY

Once the necessary safety criteria have been established, efficacy has to be established usually in Phase III clinical trials. The number of people to be included and their age, sex and geographical distribution depend on the disease involved and its prevalence in particular populations. Such trials may take a number of years effectively to determine the value of a product.

It has been suggested from time to time that vaccines originally developed for BW defense could be made available for civil use and that converted BTW facilities could have an important role in such production. However the regulatory issues involved are complex and difficult. The problems concerning facilities have already been discussed, but demonstrating efficacy is an additional hurdle to overcome.

Where any human experimentation is involved, voluntary, competent, informed and understanding consent is absolutely essential. This was one of the principles set out in the Nuremberg Code by the US judges in 1947[5,6]. A further requirement is that no experiment should be conducted where there is an a priori reason to believe that death or disabling injury will occur as a result.

Thus the efficacy of some vaccines cannot be determined in man because, even if consent could be obtained, it would be unethical and unacceptable to perform challenge tests, and in the case of some diseases field studies could not yield the necessary data on protection. Should any country have such data, they are unlikely to be made available since that would involve admitting the violation of human rights.

Human efficacy data are normally required for licensure. In the absence of this, for the reasons described above, a number of vaccines may not be licensable according to the usual criteria. It is worth noting that a number of defense vaccines developed in the USA remain subject to restrictions as Investigational New Drugs and require the completion of detailed consent forms by recipients.

Where the precise mode of action of an agent is known and data are available on immunological protection (e.g., botulinum toxin) which allows effective prediction of the level of protection by determining amounts of antibody (IgG), there is a strong case to support licensing on the basis that no human in vivo challenge is needed to confirm the efficacy. It was on this basis that a UK Product License was approved in 1991 for a botulinum toxoid vaccine. In contrast, where more complex immune mechanisms are involved, such as pertain to other bacterial, viral or rickettsial infections, simple prediction without Phase III trials is not likely to be acceptable in the present state of our knowledge of immunity.

4.3 QUALITY

As mentioned above, the detail of the production process is a vital aspect in specifying a biological product. Thus in determining if a product should be licensed all aspects of the manufacturing process will need to be considered to assess their impact on product quality and reproducibility.

4.3.1 *Seed source*

Depending on the nature of the organism, bacterial, viral, eukaryotic or recombinant, a number of tests need to be applied to confirm the origin, purity or genotype. For eukaryotic cells used to grow viruses characterization would involve isoenzyme analysis, DNA fingerprinting and cytogenetics. In the case of recombinant organisms full details of the construct need to be confirmed involving preparation of the DNA segments, genes involved, origin of replication, fusion product (if present), characterization of the flanking regions, characterization of the promoter, enhancer and terminator regions and presence of antibiotic-resistance genes.

The DNA must be analyzed by sequence analysis and restriction endonuclease digestion mapping and compared with the original gene sequence.

Master and submaster host/vector cell seed stocks will also require a similar range of tests to be applied, but in addition it is essential to monitor: (a) the physical state of the vector in the cell: (b) plasmid stability during storage, recovery and use; and (c) fidelity of the nucleotide sequence encoding the expression product, the copy number and effect of passage on that copy number.

4.3.2 *Fermentation*

The fermentation process must also be specified in terms of growth medium composition, physical parameters, final yield and limits on contaminating organisms. Equine proteins and human blood components cannot be used in culture media and the Code of Federal Regulations (CFR) requires disclosure of sources of animal-derived products used in preparing products to be administered to humans [7]. Plasmid stability may be affected by a number of factors and is a potential problem in large-scale fermentations. It can be monitored by simple screening for antibiotic resistance markers, but this does not give any indication of copy number loss.

4.3.3 *Purification or processing*

Precise details are required of the further processing after the culture stage. These will include the identity and specifications of any added reagents and any materials with which the product comes in contact, including, for example, chromatographic column matrices. Assays will be required and limits established in respect of potential carry-over of materials into the final product. Details of elution profiles from chromatographic columns, gel electrophoresis or other tests to determine purity of product are required. Other tests for potency and sterility may need to be specified at particular stages in the production process.

Various methods may need to be used to assess the identity and purity of the product, be it of 'natural' or 'recombinant' origin. In the latter instance there are very specific problems that can occur, viz--proteolytic clipping of N- or C-terminus sites, incorrect glycosylation, incorrect disulphide bond formation, amino acid substitution, oxidation, deamidation, incorrect folding after solubilization and aggregation.

The analytical methods employed to detect such deviations include: electrophorectic techniques, amino acid composition, sequence analysis, peptide mapping, nuclear magnetic resonance, circulator dichroism and mass spectrometry.

4.3.4 *Product finishing*

A licensed product would have specified the nature of the finished product (e.g., lyophilized or liquid fill) which would include details of the formulation, the container, storage conditions and shelf life. Tests on the final product would cover potency, identity, sterility (or freedom from adventitious agents), innocuity and limits on exipients.

5. Conclusion

The aim of this article has been to demonstrate that conversion of BTW facilities to be capable of the production of pharmaceutical products will be a complex and costly operation requiring careful planning and an understanding of the economic realities of making, licensing and selling medicinal products. In addition the converted facility may still be capable of reconversion because of the dual use nature of many aspects of the bioprocess industry. There are many grounds for believing that, although apparently attractive, conversion to create a pharmaceutical bioprocessing capability may not be a generally desirable conversion goal.

6. References

1. UK Medicines Act 1968 (Her Majesty's Stationery Office: London 1968).
2. United States of America Code of Federal Regulations, 21 CFR, 600.11(e)(3). 4.1.97 edn (Office of the Federal Register National Archives and Records Administration: Washington, DC).
3. Guide to Good Pharmaceutical Manufacturing Practice. (1993) Her Majesty's Stationery Office: London.
4. The Rules Governing Medicinal Products in the European Community. (1992) 4, Guide to Good Manufacturing Practice for Medicinal Products, Office for Official Publications of the European Communities, Luxembourg
5. Trials of War Criminals before the Nuremberg Military Tribuals, Tribual 1, Case 1, The Military Case: United States of America v Karl Brandt et al., (1950) 2, US Government Printing Office, Washington, DC.
6. Annas, G.J. and Grodin, M.A. (eds). (1992) The Nazi Doctors and the Nuremberg Code: Human Rights in Human Experimentation, Oxford University Press, Oxford. pp. 61-129, 188-208.
7. Cox, C.B. and Hardegree, M.C. (1976) Diphtheria and tetanus toxoids: results of assays for the presence of bovine antigens; Journal of Biological Standardization 4, 81-89.

THE POSSIBILITIES AND LIMITATIONS OF BIOLOGICAL WEAPONS CONVERSION
Personnel and Facilities

MILTON LEITENBERG
Center for International and Security Studies at Maryland
School of Public Affairs
University of Maryland
College Park, MD 20742-1811

1. The Possibilities and Limitations of Biological Weapons Conversion: Personnel and Facilities

In the United States, the two facilities for biological weapons research and production were converted between 1969 and 1972. Between the time that the Biological and Toxins Weapon Convention (BTWC) was signed and came into force-1972 to 1975--and the present time, some six to ten countries have initiated or maintained national BW R&D programs with various degrees of offensive character. None of these, however, appear to be discontinuing their programs at the present time of their own volition or under the pressure of international attention to their divergence from the BTWC. There is, therefore, no current attention being paid to questions dealing with the facilitation of conversion in these other states, Iran or Iraq for example. For the past half dozen years or a bit more, the overwhelming practical issue in the area of the conversion of former BW R&D facilities has been the BW R&D institutes of the former USSR. The examination of BW R&D conversion in this paper will therefore be built almost entirely around this particular current and continuing problem. It is also written entirely from the viewpoint of arms control; that is, economic or marketing issues are not its primary concerns, although these are among the most obvious "limitations" which will be discussed.

The former USSR's BW R&D infrastructure, as in most other areas of its defense sector, was quite large, with overcapacity in both R&D and production facilities. The exact number of institutes in question is actually unknown, at least to the international arms control community, even now, five years after the dissolution of the USSR. Paradoxically, and surprisingly, this former infrastructure appears to be "growing," as new disclosures are made. In October 1987, in its first submission of information under the BWC-CBM's the USSR reported five BW laboratories under Ministry of Defense control: (1) Leningrad (now St. Petersburg); (2) Kirov; (3) Sverdlovsk (now Ekaterinburg); (4) Zagorsk (subsequently referred to under its pre-1917 name, Sergiyev-Posad), Moscow oblast; and (5) Aralsk, Kzyl-Ordinsky oblast. The Leningrad site was under the jurisdiction of the USSR Ministry of Defense

E. Geissler et al. (eds.), Conversion of Former BTW Facilities, 119-133.
© 1998 *Kluwer Academic Publishers. Printed in the Netherlands.*

Scientific Research Institute for Military Medicine, the remaining four were under the USSR Ministry of Defense Scientific Research Institute for Microbiology. The documentation also listed fourteen civilian laboratories, for a total of nineteen sites that fit the reporting criteria, including those in Ukraine and what was then Byelorussia. The fourteen were not reported as BW facilities. However, the 1987 Soviet documentation was very incomplete. Obolensk was not mentioned in 1987, neither among the military nor the civilian institutions, but appeared in the 1989 submission as the All-Union Scientific Institute of Applied Microbiology under the jurisdiction of the USSR Ministry of the Medical and Microbiological Industry. It clearly met the criteria which should have mandated its inclusion in 1987. Then in 1992, all the disclosures about the Biopreparat system appeared, both in Russian press sources and in the West, with the result that the number of former Soviet institutions that were involved in BW R&D and production appeared to be between twenty and twenty-five.[1] In an article published in Izvestiya on October 15, 1997, a former Soviet scientist who was an advisor to this organization for several decades writes of the "birth of a system that included dozens of institutes and test ranges, institutions with the most diverse names and completely anonymous institutions, and scientific cities which denied access to outsiders." He referred to it as the "Organization P/YA A-1063" (P.O. Box A-1063), which was its early Soviet code name.[2] Later on this organization was also named Glavmikrobioprom. This was well over an order of magnitude larger than the US program had been at its height prior to 1969, and when it still included an offensive component. As troubling as the magnitude of the program was the fact that it was established by the Soviet government in 1973, and built up following that, precisely in the years during which the USSR had signed and ratified the BWC. In September 1992, then Deputy Foreign Minister Berdennikov called the Biopreparat system "one of the best-guarded secrets in the old Soviet Union," and one whose operations the Foreign Ministry professed to know nothing about as late as the end of 1992. Its management staff was apparently taken from the Ministry of Medium Machine Building, one of the former USSR's eight defense industrial ministries, and the one responsible for producing the USSR's nuclear weapons. The overall management of the entire BW R&D system appears to have been in the hands of "the 15th Directorate" of the Soviet General Staff. However, the identified facilities and their total number still appears to have been incomplete, and would apparently have included the following:

(a) the five Ministry of Defense institutes;

(b) some portion of the Biopreparat laboratories and production sites. These included the "mobilization capacity", fully outfitted, tested, and ready BW production sites, with weapon filling lines. In Russia's 1992 "Form F" submission of the Ministry of Defense, Zagorsk and Sverdlovsk are named as production sites and it is additionally stated that an unidentified number of additional "...industrial facilities with storage capabilities were created in Glavmikrobioprom... that might also produce biological agents during a crisis." One of these, an extremely large facility at Stepnogorsk, in Kazakhstan was identified by the United States in 1996. At least three other sites in Russia are also suspected of having had the same function, which would mean a minimum of six BW production facilities. These stood ready to begin production should that be ordered, until late 1991, when the weapon filling lines were removed from these locations, but the rest of the production and processing portions left intact;

(c) some or all of the 12 "Plague Institutes";

(d) some of the Academy of Sciences institutes and the 14 "civilian" R&D institutes listed in 1987;

(e) and perhaps others, some of which have recently been alleged in Kazakstan.[3]

It is possible that the majority of these institutes were primarily involved with research -- setting aside for the moment the problem of distinguishing between offensively intended research, and defensive -- and that weapon development was relegated to the Ministry of Defense laboratories and to the Biopreparat system. If that were the case, it would also be important to understand what the organizational and production status was prior to 1973. The answers to these questions are unknown.

However, the identified facilities and their total number still appears to have been incomplete. In the 1996 Russian submission to the United Nations in accordance with the annual BWTC CBMs, the "Organization of Russia's Program for R&D on Medical and Technical Measures for Defense Against Biological Weapons shows the Russian Ministry of Defense having undefined relations with fifteen institutions: four directly under the Ministry, four under Biopreparat, and seven others, including institutes primarily affiliated with the Russian Academy of Sciences, the Russian Academy of Medical Sciences, the Ministry of Health and the Ministry of Medical Products. (In its 1993 declaration, these same institutions were shown under joint organizational matrices of both the Ministry of Defense and the Ministry of Health. Therefore, exactly what these relationships are: management, funding, or contractual is not clear.) In the 1993 declaration, Russia had listed five primary facilities, presumably the five under Ministry of Defense control, and twelve others, with a total staff of at least 6,000. As to the total number of personnel that were working in the Soviet BW R&D program at its peak, the number is unknown. Zilinskas refers to "...more than 25,000 people",[4] while an article in Science in November 1997 estimated "...about 10,000."[5] Thus, the present Russian program, in 1997, still appears to be several times larger than the US program ever was, but this may be misleading as it would depend entirely on how many people within each institute were occupied wholly or in part with BW work.

At the same time, there has unquestionably been some overall retrenchment in this former system through a combination of intentional and coincidental processes. There was allegedly reported to have been some consolidation in the scope of the USSR's BW effort in the five years between 1987 and 1992. Early in 1992, General Volkogonov, at the time a spokesman for President Yeltsin, declared that "a number of centers dealing with this issue have been closed."[6] If this was in fact the case, exactly which they were, other than the Vozrozhdeniye Island test site in the Aral Sea, has never been specified. In September 1992, in conjunction with the announcement of the US-UK-Russian Trilateral Agreement, Russian officials stated that the number of personnel working in military biological programs would be cut in half, and that funding for the program would be reduced by 30 percent[7]. Given the available estimates of the number of research personnel in these former Soviet BW facilities that was not quite so great a reduction,[8] aside from the fact that there is no way to document that such reductions took place.

Nevertheless, due to the sharp drop in R&D funding by the Russian state since 1992,[9] as well as by the other independent states of the former USSR, the overall number of working scientists has dropped substantially, and that has undoubtedly occurred to varying degrees in the former BW R&D institutes as well. Even here, available estimates and numbers seem to vary substantially. In 1995, Deni and Harrington quoted "the number of research workers and science instructors in Russia-initially around 900,000 in 1991..." and that the "Russian Federation State Committee for Statistics...[noted] that between 1990 and 1993, the number of Russians employed in science and technology declined by 27 percent.[10] A 1997 report in Science News quoted a "1991...Soviet...scientific work force of 1.5 million people," with a loss of approximately 600,000, or 40 percent, of that work force by 1997.[11] The report added that as much as 80 percent of the 1991 Soviet budget for R&D was "for military projects." A Russian researcher recently used still higher figures, stating from a scientific workforce" of "about three million" in 1991, "about half" (1,5 million) remained in 1996 .[12]" This shows a greater percentage drop, but leaves as high a number in 1996 as the previous estimate began with in 1991.

Exactly how the personnel reductions are specifically reflected at the former BW institutions, whatever their actual level may have been, is not known. A Swedish scientific team visited eight biotechnology capable facilities in Russia in 1996, some of which had been part of the USSR's BW R&D infrastructure, and some of which had not.The eight facilities showed great variation, from virtually no change in employment levels to near-total extinction.[13] How representative the sample is, however, is impossible to say, and how representative it is of the former BW-related institutes, those that were directly affiliated with the Ministry of Defense, or those in the former Biopreparat system, or under any other jurisdiction, is also impossible to say. There are no known official comparative personnel levels, pre-1990 or pre-1992 and presently, for the All Union Research Institute for Molecular Biology (Koltsovo) the All Union Scientific Institute of Applied Microbiology (Obolensk), the Institute of High Purity Preparations (Leningrad), the Institute of Immunology (Lyubuchana) and others. (An excellent paper by Dr. Sandakchiev, Director of Koltsovo, follows, and an equally informative paper by his deputy, Dr. Sergey Netesov appeared in the 1994 SIPRI volume.)[14] There are also only isolated and fragmentary reports regarding the numbers of scientists emigrating. Deni and Harrington quote the figure of 9,200 scientists working who found jobs overseas in 1992 alone.[15] Most of these went to Western Europe, the United States, and Israel. The vast majority of the reduction in the number of working scientists has been due to what is termed "internal brain drain," the loss of scientists to other professions within Russia. In 1996 testimony to the US Congress, the US Defense Intelligence Agency stated that "Last year, reportedly 20-25 percent of scientists younger than 45 years old left, from one Russian BW institute alone." However, whether the "20-25 percent" were 10, 100, or 500 scientists was not stated, which would make a great deal of difference."[16] More significantly for our discussion, reliable information regarding the emigration of former Soviet-BW researchers to countries of current BW proliferation concern is totally non-existent, and what little there is are isolated, uncompiled bits of hearsay.[17] It appears however that the numbers are relatively small. At the same time there have been contacts between current Russian institutions and countries of concern.

2. New Sources of Funding for Former USSR BW R&D Institutions

Between 1994 and the present, several sources of Western grant funding became available for scientific researchers in the former USSR. Some of these programs more specifically targeted aid to researchers in former Soviet R&D institutions that had worked on weapons of mass destruction (WMD), nuclear, chemical and biological. Within this group, there was a subset even more specifically intended for researchers in former Soviet BW R&D institutes. Other grant programs did not focus on scientists within the WMD research sector specifically, but as we will see, some former BW R&D researchers benefited from these grants as well. The first of these programs, not specifically directed for the benefit of former WMD researchers, was that of the International Science Foundation (ISF), privately funded by George Soros[18] It disbursed $130 million in two years: 25,000 researchers got emergency grants of $500, and larger grants went to 3,500 joint US-FSU research teams. The ISF then ended its activities, with a small joint Soros-US government-funded successor, the Civilian Research and Development Foundation (CRDF), funded
at an initial level of $10 million for 1996. The Howard Hughes Medical Institute also provided grants in 1995 and 1996; $15 million in five-year research grants "to 90 outstanding biomedical scientists in the Balitcs, Central Europe, and former Soviet Union." Forty-six of these designated "International Research Scholars" were in the former USSR.

Research grants specifically for former USSR WMD researchers were provided
by two international and two solely US programs: the International Science and Technology Center (ISTC), the Science and Technology Center in Ukraine (STCU), and the US Industrial Partnership Program (IPP) and Lab-to-Lab program (nuclear only). Altogether, over 9,000 former weapon scientists have been funded under these programs since 1995 to "work in civilian-oriented projects."[19] Through the end of 1995, the ISTC had provided $82 million to support 202 projects; an additional 44 projects were approved for 1996 [20] Contributions to the ISTC come from the EU, the US, Japan, Sweden and Finland. Beneficiaries are scientists in Russia, Belarus, Kazakstan, Armenia, and the Kyrgyz Republic. (Further details on the ISTC program appear in the paper by Dr. Compton.) The IPP began in mid-1994 and has funded over 200 projects employing "over 2,000 NIS scientists with non-weapon related employment" as of March 1995, but their grants are largely nuclear related. The US National Aeronautics and Space Agency (NASA) has also begun funding former NIS BW R&D facilities, as will be noted in a following section.

3. The Conversion of the US BW R&D and Production Facilities: Fort Detrick and Pine Bluff Arsenal; Comparison with Events in the USSR and Russia

The US government decision in November 1969 to unilaterally end its offensive
BW program required the conversion of two major facilities, the US Army's Biological Defense Research Center at Fort Detrick, Maryland, and the Division of Biological Operations, the production facility at Pine Bluff Arsenal, Arkansas. It also required the distruction of stored agents and weapons. Britain also converted its single biological warfare R&D facility, but its size was much smaller and it had not been doing offensive work for many yeras. After the end of World War II, Britain's biological warfare laboratory was renamed the Microbiological Research Establishment (MRE). In 1979, four years after the BWC had come into force, the MRS was transferred from the Ministry of Defence to the Department of Health, and became the Centre for Applied Microbiology and Research within the Public Health Laboratory Service. At the same time, Britain's Chemical Defence Establishment (CDE) assumed responsibility for work on biological defense as well as setting up a new Defence Microbiology Division, and eventually the entire facility at Porton Down was renamed the Chemical & Biological Defence Establishment (CBDE (Dr.Pearson's paper will discuss the UK conversion experience in further detail.)

A very detailed examination of the US conversion process was published in an earlier study in 1994, and therefore only several points are included here by way of summary.[21] On May 1, 1972, the former Pine Bluff Arsenal was turned over to the Food and Drug Administration (FDA) of the Department of Health, Education and Welfare as a new National Center for Toxicological Research. The entire destruction program was carried out on schedule. Fort Detrick was divided into seveal major components, all but one of these being transferred to different civilian government agencies. One section remained as USAMRIID, the United States Army Medical Research Institute for Infectious Diseases The entire conversion process took place quite quickly, at little expense, and with minimal problems for personnel to find new professional civilian employment. Many found positions in their relevant disciplines in the facilities that took over parts of Fort Detrick and Pine Bluff, others in commercial or academic institutions. However, it is important to note major differences in the employment and economic context in the United States in 1970 to 1973 in comparison to that in the USSR and Russia from 1990 to the present time:

- The US economy was not suffering the major economic and industrial decline that the USSR and then Russia has undergone since 1990.
- The US government continued very generous funding to scientific research in general, as well as to the civilian government agencies that took over the physical plant of the converted BW installations. In contrast, the USSR and then Russia have drastically slashed R&D funding to most if not all R&D institutions in the FSU.
- Finally, US researchers were routinely accustomed to moving between positions in government, commercial or academic institutions.

4. The Basic Questions That Remain Unanswered

It is simply impossible to assess the process of BW R&D conversion in the FSU without an understanding of the most basic parameters of just what that establishment amounted to: its size, the number of institutions and researchers that made it up, and how it was organized, managed and funded. In short, one has to know:

- How many and exactly which research institutes were involved in the former USSR's BW R&D program?
- How many scientists were employed in them?
- What are each of these institutes doing now, and how many people remain employed in them?
- What was the organizations and management structure of these institutes, and how has that been altered, modified since 1992?
- Finally, is there still a covert program, as existed for twenty years between 1972 and 1992, or isn't there?

Several observations are possible following the preceeding sections. First, it is clear that it has been decided to fund R&D in former Soviet BW R&D institutes for antiproliferation reasons, without any certainty that offensive BW work has actually ended at these facilities. Obolensk and Lyubuchany work with agents for tularemia and plague. So do various of the plague institutes. Other institutes work on anthrax, smallpox, marburg, ebola, etc. Is the work "military" or "civil," "offensive" or "defensive"? If defensive, there is presumably no reason to be concerned. (On the other hand, Russian government officials currently claim that the US maintains an offensive program.) But the arms control community certainly doesn't know the answer, and if various non-Russian governments feel satisfied that they can specify these distinctions for the current Russian program, they have not provided the reasons for that certainty in any public forum. When a US-IPP delegation met with Dr. Yuri Kalinin, the director of Biopreparat, in April 1997, he refused their requests to visit Obolensk, Lyubuchany and BiokhimMashproekt, despite the fact that the delegation had letters of invitation from the directors of all three laboratories. And, quite surprisingly, Dr. Kalinin is the chair of the NASA-funded "Space Biotechnology" grant program committee. If NASA funding were to enable former Biopreparat institutes to maintain work on the same agents that they had been working on in past years, that would be exactly as undesirable from an arms control and conversion viewpoint as if the program were nominally one to support "dual threat" agents. A Swedish expert felt that no funds should be directed to privatized companies located on the facilities of former Soviet BW institutes. This is a legitimate arms control question, and not one solely relegated to uninformed or "anti-Russian" politicians. Poland had and currently maintains the Military Institute of Hygiene and Epidemiology, which has engaged in research on botulinum toxin and anthrax for decades. What was the nature of this work in the past and what has it been since 1989-90? The answers are not in the public record.

5. The Limitations

There unquestionably are major limitations to BW R&D conversion in present-day Russia, but they are all without exception "non-specific" to the scientific questions of conversion, that is, the problems of research capability or applicability of scientific staff or of laboratory equipment. They are instead grounded in structural problems of the Russian economy, or government funding and policy.

(1) The general state of the Russian economy, the low earning levels of the great majority of "blue collar" workers, the low remittance levels of retired pensioners, etc.

In contrast, the newly-rich strata in Russia prefers to pay for Western pharmaceutical and medication when they require them.

(2) The incompetence of the Russian government in managing the federal budget, in particular, the enormous deficit in revenue-gathering and taxation, especially from corporate entities. This results in the inability to allocate funds for--among other purposes--research laboratories that depended on federal funds for their budgets, to biotechnology institutions, to former BW R&D facilities for conversion, etc. In 1997, the Russian government raised only half the revenue by tax collection that its treasury anticipated, which caused an extended delay in the disbursement of IMF loans committed to Russia, and early 1998, the continued shortfall of tax gathering prevented the Russian government once again from carrying out its promises to pay back-salary to state-supported research institutes [22]

(3) Continuing federal, regional, municipal, local impediments to foreign investment and joint ventures, including constantly changing legal and taxation parameters, as well as various forms of corruption. (Example will be provided) In 1995, Deni and Harrington wrote, 'Private investment in the NIS remains treacherous, as confusion over federal and local tax laws, import duties, and intellectual property rights continue to plague international investors trying to do business in the NIS." There apparently has been no marked improvement in these sorts of complications since 1995. Deni and Harrington (?) also reported that Western pharmaceutical producers have used bribery to accelerate obtaining licensing for the sale of their products in Russia, thus undercutting the competitive abilities of competing locally manufactured medications in Russia.

(4) Resistance to conversion. This takes two forms: policies of the Russian government and the military command structure of the 'Former Soviet Union BW R&D Complex" (FSU BW R&D Cx) in particular, and of the directors of specific institutes that were part of that complex. It appears that with the passage of a half dozen years of restrictive and declining Russian federal R&D budget allocations, and the arrival in 1994-1995 of Western grant funds, there has been a reduction if not yet a disappearance of resistance to conversion on the part of at least some of the directors of FSU BW-related laboratories. It is the other half of this equation that is of continuing and even increasing concern. There are repeated indications that the Yeltsin government still does not have control over the entire FSU BW R&D Cx. The alternative explanation of events over the last three or four years may be even worse; that the Yeltsin government has decided to obstruct transparency and the ability of other states to be certain that Russia's offensive BW program has ended, and that Russia is in compliance with the BWTC. Russia has frozen the US-UK-Russia trilateral process of laboratory site visits for four years. It has refused access to Russia's Ministry of Defense BW laboratories, which was an integral part of the 1992 Trilateral Agreement, and it has refused return visits to the Biopreparat laboratories since the

initial round of visits in 1993, although there have been repeated visits to several of these outside of the Trilateral Framework. Russian claims that they do not want to continue the trilateral inspections because the US maintains an offensive BW program are not credible. If Russian officials believed that to be the case, then they should be strenuously demanding expanded and routine access to US facilities in the course of a bilateral or trilateral process. Instead, they have frozen all inter-governmental site visits.

The Russian government's positions in the effort to develop a verification protocol to the BWTC in the Ad Hoc Group can certainly not be considered accidental or outside its control in any way. They are presumably elaborated at the Russian Foreign Ministry, although the role of the Ministry of Defense may also be significant. Unfortunately, they are consistent with and reinforce the retrograde Russian policies in the context of the trilateral process. Russia has insisted on the adoption of two measures that it knows are entirely unacceptable to Western governments. The first is to list the agents that would be included under the BWC's Article I prohibition, and the second is to define threshold limits for allowable quantities of BW agents. Further, Russian negotiators have gone so far as to privately inform other negotiators that they hope that the effort to develop a verification protocol will fail. Even something that could otherwise be considered derisive, but at the same time trivial, takes on extremely negative implications. Russian negotiators in the Ad Hoc Group in Geneva have returned to the pre-1992 "construction" of events regarding the outbreak of anthrax in 1979 in Sverdlovsk (as Ekaterinburg was named at the time). Unless there can be trust that Russia's BW program is purely defensive, external private investment is likely to end, external grant funding will certainly end, and "conversion" will patently not be taking place, even if some reductions in the overall magnitude of the program take place. Most important of all, damaging effects will spread to other areas of US-Russian WMD arms control in the most shocking, damaging and explosive manner imaginable.

6. Alternative Research Areas

This question has been examined for over 30 years. Focusing solely on research capabilities and omitting economic issues -- the source of funding, either government or private, and the marketability of commercial products -- the answers are no different than they were previously. Above all else and most certainly of all is the conclusion that maintaining work on the category of "dual-threat agents" is the least desirable of all possible alternatives. It provides little or no conversion at all. It would maintain researchers working with the same BW agents that they had been working with before and doing the same work that they had been doing previously, at the same time as it would supply a new rationale and guise for their activities. The latter could even be seen as a perverse outcome of such a suggestion, and it is highly undesirable. Moreover, it is scarcely as if there were nowhere else to turn to in seeking alternative research paths.

7. Focus on Developing Nations

If the desired focus is providing benefits to the health of populations in the developing nations through the conversion of formerly military R&D facilities in the biological or microbiological area, as early as 1963 a conference on the Global Impacts of Applied Microbiology took place in Stockholm,[23] and in the same year, the United Nations Conference on Science and Technology made available an inventory of the benefits that different scientific and technological disciplines could bring to bear on the problems of developing nations. These were repeated in several of the ACAST reports -- the Advisory Committee on Science and Technology in Development of the UN- produced in the early 1970's. In 1979, SIPRI published The Fight Against Infectious Diseases: A Role for Applied Microbiology in Military Redeployment.[24] The study did not attempt to establish priorities for redeployment, but that is easily and convincingly provided by other sources. One has only to look at two tables produced each year by the World Health Organization (WHO): (1) The World's Most Prevalent Diseases, which tabulates the number of people infected globally, and (2) The World's Most Common Cause of Death from Disease. Each table begins with a category of "vaccine preventable diseases." Other statistical compilations which facilitate the same selection are the WHO's "Infection-Associated Child Mortality," and data on the incidence of tropical and parasitic diseases. In 1993, 267 million people had malaria. (In 1994, the WHO indicated that between 275 and 490 million people suffered from malaria.) Two hundred million had schistosomiasis, about 100 million had leprosy, 89 million lymphatic filariasis, 17 million river blindness, 17 million Chagas disease. In 1990, 1.7 billion people were infected with tuberculosis, 49 million with measles, and 46 million with pertussis. Nearly 0 percent of the world's population lives in regions where malaria is endemic, and the disease is currently spreading into areas that have been malaria-free for decades. Mosquitoes transmit disease to at least 700 million people each year, about one-eighth of the world's population, and cause about 3 million fatalities per year. Malaria alone is responsible for 25 percent of childhood deaths in the third world. A consortium of the world's leading pharmaceutical companies recently rejected a proposal to join public and private organizations in research on malaria and other tropical diseases.[25] At the same time, a major portion of present US government research for a potential malarial vaccines is funded through the medical laboratories of the US Navy and US Army [26] Obviously, this should therefore be a prime candidate for US government funding for FSU BW R&D institutes.

On top of all these, there is the rapidly expanding catalogue of new pathogenic agents recognized since the early 1970's -- the statistics on the global incidence of AIDS being only one prominent example -- the reemergent infections in the last two decades, and the global spread of antibiotic resistance in a wide variety of human pathogens.[27] In the face of the numbers quoted above, as well as these other broad areas of particular concern, work on "dual threat" agents should be at the very bottom of any priority list based on relative utility. Then there is the additional consideration that work on such agents essentially means maintenance of BW capability and non-conversion, and that second factor should rule them out of consideration entirely. Thirty years after convening the 1963 conference in Stockholm, and nearly twenty years after his SIPRI volume, Carl Göran Hydén again commented on the conversion of BW R&D capabilities ("redeployment" in his phrase) in a remarklable paper od nearly prescient insight. He specified areas of work beneficial to developing nations as "... methods to upgrade foods, practice biological pest control, recycle organic materials, improve soil structure, detoxify chemicals, and improve recovery of mineral resources"[28] The clear and absolute avoidance of anything having to do with "dual threat" agents is obvious.

More recently, and for exactly the same reasons, a proposal in November 1997 by a committee of the US National Academy of Sciences (NAS) suggesting that projects dealing in particular with anthrax, plague, viral hemorrhagic fevers, and two other
directly BW-relevant agents be funded in former BW R&D institutes in Russia is an extremely unfortunate mistake.[29] A simple question to the Director and the Deputy Director, respectively, of two of the relevant institutes in Russia immediately obtained the response that of course they would have no objection to obtaining research funds to work on malaria, or other third world parasitical diseases that kill millions annually. But in any case a large number of research problems in molecular biology that do not deal with disease agents at all could have been specified, problems which require the most advanced competence in molecular genetics, as well as having direct applied and commerical possibilities. The remark of Dr. Peter Jahrling, as a justification for the NAS panel's choices, that "Any one of these projects that doesn't entail looking at Paramecia in pond water has a certain level of risk" is simply not true. The November 1997 issue of Scientific American contained in the single issue articles on:

- Using genetic engineering to protect rice from disease. The author pointed out that when she began her work on this problem in 1990, no one had ever succeeded in cloning a disease-resistant gene from any plant. The blight resistant segment of the rice DNA was cloned in bacteria.[30] The beginning of field trials of human antibodies produced by field crops -- corn in this case -- against tumor cells; it is hoped that the plant antibody will avoid human rejection caused by cloning done in mouse cells [31] If a single issue of Scientific American can provide two such examples of the highest caliber molecular genetics -- both with obvious commercial application - then there should be no problem for the members of an NAS panel or the molecular geneticists at any of Russia's former BW R&D institutes to come up with equally valid research topics to make use of their talent that are very far from 'Paramecia in pond water," and equally far from BW relevant organisms. As if to emphasize this with pointed irony, a report in the general press in December 1997 described the advanced DNA research being carried out on Oxytricha, a protozoan cousin of 'Paramecia in pond water.'Far more the point, research on developing a malaria vaccine, stressed earlier, has recently focussed on DNA vaccines, one of the most advanced techniques of molecular biology [32]

8. Focus on Russia and the Former USSR

This could be examined in several ways:

- through a survey of statements made in the early 1990's by former Soviet BW R&D institutions and their management regarding their conversion interests and capabilities
- through an examination of what has happened in various of these institutions in the past half dozen years through,
  - local production initiatives
  - commercial joint ventures with foreign pharmaceutical corporations
  - funding for research provided through external grant funders: The International Science and Technology Center (ISTC) and the Science and Technology Center in Ukraine (STCU), and the International Science Foundation (Soros).

Other papers in this compilation examine the ISTC grant program, as well as the conversion of the Stepnogorsk BW production facility. Broadly speaking, some of the major areas of commercially valuable microbiological research that could be engaged in are the following:

(1) drugs and pharmaceutical products
(2) remediation of oil pollution
(3) remediation of other environmental contamination (heavy metals, etc.)
(4) microbial insecticides
(5) microbial herbicides
(6) anti-food spoilage
(7) anti-industrial and raw material microbial destruction
(8) malarial, TB, or other vaccines for diseases on the WHO list
(9) nitrogen fixation of non-leguminous food crops, particularly grains and corn.

(Regarding this last, success with bacterial inoculation of non-leguminous crops was reported in the USSR by the mid-1960's. These reports may not have been valid, nevertheless this would be an area of research of inestimable value to global agricultural production, agricultural economics, and the continued viability of both agriculture and the earth which sustains it.) An earlier study again contained a detailed catalog of Soviet and Russian statements between 1990 and 1994 on capabilities, intentions, and plans for conversion of BW facilities and research. They are therefore not repeated here.[33] Biological warfare R&D laboratories that formerly produced vaccines -- the Ekaterinburg laboratory was one of these -- could also produce vaccines for domestic livestock; cattle, hogs and poultry are all routinely vaccinated against various diseases. Other projects could also be applied to the production of protein feed and fertilizer utilizing bacterial cultures which transform paper waste to cattle feed or to fertilizer. Pilot plant installations for these were developed as long ago as the mid-1960s in various US universities. For each of the sites a list of between five and ten alternative research or production priorities should be drawn up. These would fall into two broad categories: (a) the production of commodities and services which the institutes by their nature can produce and may already have produced in the preceding years, and which are in the most dire need in Russia and in the other new republics of the former USSR, and (b) interdisciplinary problem-oriented research in the fields of public health and ecological and environmental problems.

The entire territory of the former USSR -- as well as most of Eastern Europe -- suffers from gross pollution in every category: air, water, land and, over large areas, also radiation.[34] Any products these institutions could produce that served environmental protection or recovery would also have immediate application.

9. References

1. Leitenberg,M. (1994) The Conversion of Biological Warfare Research and Development Facilities to Peaceful Uses, in E. Geissler and J.P. Woodall (eds.), *Control of Dual Threat Agents: The Vaccines for Peace Programme,* Stockholm International Peace Research Institute and Oxford University Press, . 77-105, and Leitenberg, M. (1996) Biological Weapons Arms Control, *PRAC Paper no. **16**, Center for International and Security Studies, University of Maryland,*
Rimmington,A. (1996) From Military to Industrial Complex?, *Contemporary Security Policy, **17**,* 80-112.; Caudle,L.C. (1997), The Biological Warfare Threat in F.R. Sidell et al. (eds.) *Medical Aspects of Chemical and Biological Warfare, Washigton, DC, Office of the Surgeon General,* 454-455.

2.Yankulin, V. (1997), Plague Syndrome, or the Purgatory of One of the Creators of Bacteriological Warfare, *Izvestiya,* October 15, .

3. In November 1996, Dr. Genady Lepeshkin, the director of the National Center for Biotechnology in Kazakstan, indicated that there were nine biotechnology institutes (in Kazakstan), all of which had been "part of the defense complex." Apparently only one of these, Stepnogorsk, had previously been identified.
It appears that one other of these was part of the Biopreparat organization; the remaining seven were under the authority of other Soviet ministries and agencies.

4. Zilinskas, R.A.,(1997) The Other Biological Weapons Worry, *New York Times, November 28.*

5. Stone,R., (1997) An Antidote to Bioproliferation, *Science, **278*** , p. 1222. Stone however adds that "...an estimated 150 to 200 scientists still work on biological defense related projects at four centers", which seems to remove any credibility to his estimates. If that were the case, 9,800 researchers would already have been "converted" in one way or another.

6. Devroy, A and Smith,R.J.(1992) , U.S., Russia Pledge New Partnership, *Washington Post, February 2.*

7.Dahlburg, J. T. (1992), Russia Admits Violating Biological Warfare Pact, *Los Angeles Times,* September 15, and Gordon, M.R. (1992). Russia and West Reach Accord on Monitoring Germ Weapon Ban, *Washington Post, September 15.*

8.Leitenberg, M. (1994) The Conversion of Biological Warfare," op. cit., and Leitenberg,M.(1994) Biological Weapons Arms Control, (note 1)

9. For overall surveys, see:
Leonid Kosals, "Military R&D Institutes in the Context of Demilitarization in Russia," WP-94-002, International Institute for Applied Systems Analysis, Laxenburg, Austria, January 1994.
Ksenia Gonchar, Research and Development (R&D) Conversion in Russia, Report no. 10, Bonn International Center for Conversion, Bonn, Germany, May 1997.

10.Deni , J. R. and Harrington A. M. (1995) Beyond Brain Drain: The Future of Nonproliferation Through Science Cooperation' Programs, *Conference Paper,* March *C30-31 Center for International and Security Studies, University of Maryland, quoting FBIS-SOV-93-067, April 9, 1993, p. 39, and Izvestia, April 30, 1993, p. 15, in FBIS-SOV-93-085, p. 40.*

11.Vergano, D., (1997) Rough Times in Russia: Post-Soviet Science Faces a New Crisis, *Science News, May 10.*

12. Zilinskas,R. personal communication, October 1997.
(USSR Science: Novosti Press Agency, 1973) published the following data for the USSR:
1940: 2,359 research centers (colleges and universities included), including 786 research institutes and their local branches;
1960: 4,196 scientific centers, with a research staff of 354,200;
by the end of 1970: 4,985, including 2,458 research centers and their branches, and 927,700 (research staff);
1973: over 5,300 research establishments and one million researchers.

13.Öhrner, N. et al., (1996) Some Views on Russian Biotechnology in Academic Institutions, State Research Centres, and Industry,*Report from IVA's Biotechnology Delegation to Moscow and Obolensk, 11/16-22/96.* It should be noted that there is no external verification of these estimates.

14.Netesov, S. (1994) The Scientific and Production Association Vector: the current situation" in: E. Geissler and Jack P. Woodall (eds.), p.133-138, note1

15. Deni and Harrington, 1995, op. cit., p. 7.

16. Current and Projected National Security Threats to the United States and its Interests Abroad, Hearing before the Select Committee on Intelligence, United States Senate, 104th Congress, 2nd Session, February 22, 1996, Washington, DC: US Govt. Printing Office, 1996, p. 198.

17. There have been suggestions that Iran, Libya and Syria may have obtained aid for their BW programs from the former USSR. Most of the indications concern Iran more recently, in the post-1992 years. There are also indications that the USSR supplied some assistance to Iraq's BW program before 1990, as did a sizable number of West European countries. See M. Leitenberg, Biological Weapons Arms Control, op. cit.

18. International Science Foundation, Annual Report 1995, and 1994 Annual Report Supplement, New York, New York.

19 Deni and Harrington, op. cit., p. 17.

20 Zilinskas,R.(1997) personal communication.

21 Leitenberg, M. (1994) The Conversion of Biological Warfare Research and Development Facilities to Peaceful Uses in E. Geissler and J.P. Woodall (eds.) Vol. **15** *SIPRI Chemical and Biological Warfare Studies* , Oxford University Press, pp. 77 - 108 . It is important to indicate that this study of the dismantling of Fort Detrick and Pine Bluff was originally prepared in 1984 for a United Nations Secretary General's study on Military Research and Development, and together with a second detailed study prepared by a Fort Detrick staff member, both became directly available to the Soviet government. This is of importance because Soviet officials claimed that they did not believe that the US had dismantled the two offensive portions of its program.

22.Williams, D.(1998) Russia Falls Short on Promise to Pay Back Wages, *Washington Post,* January 8.

23.Hedén,C.G.(1967) Microbiology in World Affairs", *Impact (UNESCO),* **17**, 187-208. See also: The Resolution of the Second International Conference on Global Impacts of Applied Microbiology, Addis Ababa, November 1967.

24Hedén C.G. and Sinyak,K.(1979) The Fight Against Infectious Diseases: A Role for Applied Microbiology in Military Redeployment, *SIPRI, Stockholm International Peace Research Institute,*

25.Greenberg, D.A. (1998) A Pittance to Fight Malaria, *Washington Post,* January 4.
Williams, N. (1997) Tropical Diseases: Drug Companies to Collaborate, *Science* ***278***, 1704

26.Hoffman, St., Briefing on United States Deparetment of Defense Malaria Research Programs. naval Medical Research Institute, Malaria Program, May 14, 1996. The irony is that the total US government budget for malaria research is on th order of $ 25 million per year, with malaria a major medical concern for US military forces, while the US Congress mandates that the US Army spend $ 135 million per year on breast cancer research, which should not be considered within the purview of military medical concerns at all.

27.Infectious Disease - A Global Health Threat; Report of the National Science and Technology Council, Washington, DC, September 1995. See also: Anne E. Platt, Infecting Ourselves: How Environmental and Social Disruptions Trigger Disease, World Watch Paper No. 129, April 1996, Washington DC.

28.Hydén,C.G. (1992) The 1991 Persian Gulf War: Implications for Biological Arms Control in Zilinskas,R. (ed.) The Microbiologist and Biological Defense Research: Ethics, politics, and International Security, *Annals of the New York Academy of Siences, Vol.* ***666***, *1-8*

29.Stone, R. (1997) Russia: An Antidote to Bioproliferation, *Science,* ***278***, p. 1222.

30. Pamela C. Ronald, "Making Rice Disease-Resistant", Scientific American, 277:5 (November 1997), pp. 100-105.

31. Gibbs, W.W. (1997) Biotechnology: Human Antibodies Produced by Field Crops Enter Clinical Trials, *Scientific American, 277,* 44.

32.Sawyer, K. (1997) With DNA, Simple Cells Perform Acrobatic Feats of Computation", *Washington Post, December 15,* Tauber,G.(1997) Salvation in a Snippet of DNA? *Science* ***278***,1711-1714, and Ebola Gene Vaccine is promising in Animals, New York Times, December 30, 1997.

33. Leitenberg, M. (194) The Conversion of Biological Warfare Research and Development Facilities to Peaceful Uses," ,note 1

34.Feshbach M. and Friendly, A.,Jr., (1992) Ecocide in the USSR: Health and Nature Under Siege, Bloomington Indiana: Basic Books 1992; see also Shapiro, M. (1992) Ex-Soviet Empire the New 'Sick Man of Europe', *Washington Post, October 3.*

CONTROLLING DANGEROUS PATHOGENS: A BLUEPRINT FOR U.S.-RUSSIAN COOPERATION

A Report of the U.S. National Academy of Sciences to the Cooperative Threat Reduction Program of the U.S. Department of Defense

CHRISTOPHER P. HOWSON
Institute of Medicine, National Academy of Sciences
2101 Constitution Avenue, N.W., Washington, DC 20418, USA

1. Introduction

The National Academy of Sciences (NAS) created the Committee on International Security and Arms Control (CISAC) in 1980 to bring the scientific and technical resources of the Academy to bear on urgent problems of international peace and security. The primary initial activity of CISAC was a dialogue with Soviet counterparts that helped keep communication on nuclear arms control issues open during a time of great tension in U.S.-Soviet relations. In 1986, CISAC created a special working group on biological weapons control, which focused on another critical problem—developing improved methods of verification of the Biological Weapons Convention (BWC). The group carried out bilateral discussions with a counterpart group established under the aegis of the Soviet Academy of Sciences and then supported by the Russian Academy of Sciences.

Beginning in 1993 the working group became particularly concerned about potential proliferation of biological weapons (BW) capabilities because of the economic difficulties that afflicted the former Soviet BW complex, along with other Russian institutions, after the collapse of the Soviet Union. At the same time, the group was impressed with the Russian expertise in the biological sciences and biotechnology. There appeared to be a good opportunity to draw on that experience in a cooperative effort to combat the global threat of emerging infectious diseases and promote U.S. national security interests. Members of the working group began to discuss this opportunity with appropriate officials of the U.S. government and encouraged efforts such as the decision of the International Science and Technology Center (ISTC) to fund appropriate research projects at former Soviet BW research facilities. In 1995 the U.S. Department of Defense (DOD) asked CISAC for assistance in designing a program to expand bilateral cooperative efforts between U.S. scientists and their Russian counterparts who had been involved in the former Soviet BW program.

The project began in the fall of 1996, with funding from the Cooperative Threat Reduction (CTR) Program, commonly referred to as the Nunn-Lugar Initiative. DOD provided CTR funds for supporting pilot research projects at Russian institutes to examine the potential for collaborative research activities that could be carried out

E. Geissler et al. (eds.), Conversion of Former BTW Facilities, 135-148.
© 1998 *Kluwer Academic Publishers. Printed in the Netherlands.*

effectively at facilities involved in the former Soviet BW program. A 14-member committee, which included members of the CISAC working group, the co-chair of the Institute of Medicine's Board on International Health, and additional experts on BW and international health issues, developed the plan presented in this report with assistance from staff. Appendix A contains the roster of committee members and staff.

The committee's report and its recommendations were submitted to DOD in October 1997 where they are currently under review. The recommendations are intended to complement other U.S. government activities supporting U.S.-Russian cooperation and to provide a conceptual basis for future multinational collaboration on dangerous pathogens. This paper summarizes the major elements of the NAS report.

2. The Importance of U.S.-Russian Cooperation on Dangerous Pathogens

Russia will undoubtedly continue to support legitimate research and related activities on dangerous pathogens.[1] U.S. involvement in these activities through cooperative programs will provide opportunities to build confidence that prohibited research is not being pursued under the guise of legitimate undertakings. Also, from a public health viewpoint, Russian scientists who participated in the Soviet BW-related activities have unique knowledge of many aspects of naturally occurring pathogens as well as those that could be used by terrorist groups.

The committee believes that appropriately structured U.S.-Russian cooperation on dangerous pathogens—featuring direct laboratory-to-laboratory contacts and based on the principle of broad transparency—will serve the interests of both countries. Such cooperation could contribute significantly to the following objectives:

2.1. NATIONAL SECURITY BENEFITS

<u>Providing greater mutual confidence about compliance with the BWC than would otherwise be possible</u>: Properly designed collaborative research projects can provide an important means for enhancing transparency: joint research, person-to-person contacts, regular exchanges of personnel, and direct access to facilities all promote transparency. Many defense scientists working in closed facilities have had little contact with civilian counterparts working on related problems. Involving these scientists in research on public health problems could build bridges between military and civilian institutions and their personnel. Integrating them into international, as well as national, networks of researchers committed to the prevention and control of dangerous diseases should reinforce standards of ethics and social responsibility that counter the temptations of illegal activities. Providing opportunities for them to talk shop with colleagues and, as a result, to practice their trade and be valued for their contributions are strong incentives for complying with international norms.

---

1 For the purposes of this report, dangerous pathogens are defined as pathogens that are highly infectious, causing great concern to global public health. Of particular interest are pathogens that could be used in biological warfare.

Reducing proliferation incentives: For several years, Russian scientists who possess extensive experience in handling pathogens with BW potential have left their institutes for new careers that promise better financial or scientific futures in Europe, the United States, and Israel. Some of the remaining scientists may be tempting targets for states or terrorist groups in search of recruits or information on BW. Collaborative programs, with guaranteed paychecks and challenging research activities, can reduce the economic incentive for Russian specialists to respond to such recruitment efforts.

Serving as a stepping stone to dismantlement opportunities: Collaboration on dangerous pathogens can help identify opportunities for joint dismantlement projects, another key CTR objective. Experienced researchers can provide guidance on the most effective use of facilities and on the removal or modification of research equipment no longer needed for military purposes. Research institutes also might provide introductions to other parts of the former BW complex, such as engineering and design facilities, where excess military-oriented equipment could be put to new types of peaceful use.

Reconfiguring former Soviet military-related activities: Severe cutbacks in funding for military programs have already led to a considerable redirection toward civilian activities of research and development that previously supported the Soviet BW program. However, large and diffuse research and development and standby production capabilities with BW potential remain in Russia. In the absence of alternative employment opportunities, defense scientists constitute a powerful lobby for maintaining facilities that are beyond Russia's national security requirements for defensive BW research. A substantial collaborative research program would provide new employment opportunities for many key scientists, thereby reducing pressures on the Russian government to invest in maintaining unnecessary facilities. Also, if such a collaborative program is designed to concentrate financial support at a limited number of the best Russian facilities, any weaker facilities—to the extent they remain viable—will have incentives to find work outside the area of dangerous pathogens, particularly in the private sector.

Enhancing capabilities to combat bioterrorism: Both the United States and Russia are concerned about the growing threat of bioterrorism. Expanded cooperation in basic research, epidemiology, diagnosis, and prophylaxis of diseases associated with dangerous pathogens can enhance the capabilities of both countries to identify and respond to emerging terrorist threats.

2.2 PUBLIC HEALTH BENEFITS

Improving understanding of the prevalence and characteristics of pathogens that pose threats to public health: The Soviet investment in BW-related research has resulted in a cadre of highly qualified scientists with unique knowledge about dangerous pathogens. Collaborative activity can provide the United States with new insights into research capabilities, laboratory techniques, and knowledge that may not currently be known outside Russia.

Strengthening capabilities to prevent, diagnose, and treat outbreaks of infectious diseases: Both countries have substantial capabilities in epidemiology and public health for monitoring outbreaks of diseases. Linking national capabilities through expanded

cooperation can enhance global surveillance and improve epidemiologic investigations and responses to outbreaks of diseases.
Enhancing international communications concerning disease trends and outbreaks: Strengthened communications links among individual investigators, research institutions, and government agencies can improve the capabilities of both countries to anticipate and respond to outbreaks of diseases and provide mechanisms to determine the source of outbreaks.

2.3 ECONOMIC BENEFITS

Improving the stability of Russian research institutes by increasing the commercial viability of their research products: Defense scientists with BW expertise are more likely to continue working on projects with promising economic futures, both for their institutes and for themselves, than to remain in place if they are offered only low-paying tasks with little future or very short-term technical challenges with uncertain long-term security. Each research product that achieves commercial viability is also a small contribution to Russia's transition to a market economy. At the same time, the reality is that Russian institutes have had great difficulty finding international or domestic markets for their products or finding partners who can help locate such markets. Some of the projects included in the collaborative program should help a few Russian institutes become more competitive commercially by focusing new attention on the importance of high-quality prototypes, strong quality-control systems, and well-developed marketing and distribution systems. Cooperation will provide opportunities for access to U.S. business and marketing skills in these and other areas.
Leveraging limited national financial and human resources to serve national security and public health interests: Coordinated research activities in both countries on dangerous pathogens that pose current and potential health risks to the world's populations offer opportunities to combine limited financial and human resources in combating dangerous infectious diseases.
Providing new opportunities for the U.S. private sector to become more active in Russia: The current level of U.S. private-sector involvement in Russia in the development of vaccines, diagnostic devices, or other commercial commodities in the biomedical field is very low. Cooperation can lead to better appreciation of the capabilities of Russian specialists and provide easier access to Russian expertise and facilities. Coupled with appropriate commitments to respect intellectual property rights, such developments can increase the interest of the U.S. business community in investment, licensing arrangements, and expanded trade.

2.4. SCIENTIFIC BENEFITS

Enhancing the base of fundamental knowledge of pathogenesis: A collaborative effort that engages Russian and U.S. scientists in fundamental science will provide opportunities for exploring new research avenues.
Increasing the international availability of research results: Most Russian scientists cannot afford to subscribe to Western journals or attend scientific meetings in the West. On the other hand, U.S. scientists have a limited appreciation of the past

accomplishments of Russian investigators because of a lack of English-language reports of their scientific findings. The proposed program to engage key Russian specialists in joint projects should significantly increase the exchange of information and knowledge between the two countries.

Even the most extensive collaboration between Russian and U.S. scientists will not provide incontrovertible assurance that all research activities on dangerous pathogens are devoted to legitimate purposes. Recognizing this risk, the committee has concluded that, if governed by appropriate rules of transparency, a cooperative program can be carried out in a manner that ensures the risk of abuse of such a program is reduced to an acceptable level.

The expanded arrangements to provide sufficient transparency should include mutual agreement on a project-by-project basis of activities that are legitimate; regular and agreed-upon access to facilities, personnel, and information; and commitment to the principle that providing assurance is an active rather than a passive responsibility. Moreover, during the evaluation of any joint research project, its potential contributions to health and national security must be judged to outweigh the risk that the project might contribute to development or improvement of offensive BW capabilities.

The committee recommends that the proposed program be undertaken in close coordination with related bilateral activities (see Table 1).

3. Establishing the basis for expanded cooperation

The committee's consultations with a range of Russian officials, research managers, and laboratory scientists helped ensure that assessments of the technical basis for cooperation were authoritative and realistic. In addition, these interactions resulted in Russian specialists acquiring a sense of genuine partnership in development of the recommended program. Consultations included scientific visits to Russian research institutes in Koltsovo and Obolensk, where pilot projects were later established; one round of consultations and two joint planning meetings in Moscow; and an international symposium in the Kirov region involving 30 Russian specialists, sponsored by the National Academy of Sciences (NAS) and the International Science and Technology Center (ISTC).

Of particular importance is the reported endorsement by the Russian Defense Council of the NAS initiative. Such support would be critical to the future success of a cooperative program that would involve defense scientists.

TABLE 1 Selected Organizations with Program Interests Related to the *Pathogens Initiative*

Organization	Description
International Science and Technology Center (ISTC)	This international organization, established by the United States, European Union, Japan, and Russia, has supported a number of projects at a variety of institutes that were involved in the former Soviet BW program, including projects that address dangerous pathogens. ISTC has received dozens of other well developed, but yet-to-be-funded proposals in this area. In addition, it has sponsored several symposia on related topics.
Initiatives for Proliferation Prevention (IPP)	This program of the U.S. Department of Energy recently announced its intention to support biotechnology proposals at Russian institutes, particularly Biopreparat institutes, which have potential for commercial markets. IPP has received dozens of proposals from a variety of Russian institutes, including proposals for research on dangerous pathogens.
U.S. Civilian Research and Development Foundation (CRDF)	This private foundation, established by Congress in 1992 and set up by the National Science Foundation in 1995, has solicited basic research proposals from interested Russian investigators in the biomedical field. CRDF currently supports eight projects at institutes that were part of the former Soviet BW complex. The future of additional competitions is subject to further funding.
National Aeronautics and Space Administration (NASA)	NASA has funded several projects at Biopreparat institutes in support of its space science activities. The future of these types of projects in Russia is uncertain.
Centers for Disease Control and Prevention (CDC)	The CDC has long-standing relationships with several Russian institutes, some of which participated in the former Soviet BW program. It has participated in activities in Russia supported by the U.S. Agency for International Development and frequently provides training for Russian scientists.
National Institutes of Health (NIH)	NIH provides grant supplements to its U.S. investigators to involve Russian scientists in their projects. NIH also provides training opportunities for Russian specialists in its laboratories. Russian researchers who participated in former Soviet BW activities are included in these programs.
U.S. Department of Agriculture (USDA)	The USDA supports a limited number of biotechnology projects in Russia that are directly linked to agriculture development. Its current Russian portfolio does not include research on dangerous pathogens.

Biopreparat—an organization originally established by the Soviet military establishment to provide a wide range of BW-related research, production, and support services—was the principal point of contact in Russia for the NAS committee. Specialists from many Biopreparat facilities and other Russian organizations participated in these consultations.

The NAS committee was not successful in its efforts to meet with specialists from the Russian Ministry of Defense (MOD), but efforts to engage the MOD should continue. Several Russian officials have expressed optimism that the MOD will eventually participate in bilateral cooperative activities, either directly or in partnership with Biopreparat institutes. Although MOD participation is highly desirable, the Biopreparat complex provided much of the critical research and development support for the Soviet program, and thus the committee believes that cooperation with Biopreparat in and of itself will make a valuable contribution to achieving the benefits mentioned above.

In parallel with these consultations, the NAS committee initiated six pilot projects at two Russian facilities—the State Research Center for Virology and Biotechnology "Vector," in Koltsovo, Novosibirsk Region, and the State Research Center for Applied Microbiology, Obolensk—and is developing two more (see Table 2) to gain experience conducting collaborative research projects. The institutes at Koltsovo and Obolensk were chosen for four reasons: 1) as a result of the important roles the two institutes played in the BW program of the former Soviet Union, they have some of the strongest research capabilities related to dangerous pathogens in the country; 2) these institutes have strong linkages with other institutes of the Biopreparat complex, institutes of the MOD, and other institutes with capabilities related to dangerous pathogens. Therefore, they provide good initial points of entry in development of a program of national scope; 3) through previous NAS activities in Russia, NAS committee members were personally acquainted with the directors and other personnel of the two institutes and believed that they would be receptive to projects involving active bilateral collaborators; and 4) ISTC had been successful in initiating a few projects at the two institutes, beginning in 1994, indicating that these institutes were prepared to overcome administrative hurdles in developing cooperative activities and would agree to provide access to their facilities.

The pilot projects have contributed useful insights at the scientist-to-scientist level about the capabilities of the two countries. The projects also were important in convincing Russian colleagues that the NAS undertaking was a serious endeavor with strong backing from the U.S. government, thereby encouraging them to participate actively in planning a long-term program.

Table 2 Pilot Projects Initiated by NAS and Financed by DOD

Projects under way as of July 1997, with funds committed to Russian institutions[2]

At the State Research Center for Virology and Biotechnology "Vector," Koltsovo

ℵ Study of prevalence, genotype distribution, and molecular variability of isolates of hepatitis C virus in the Asian part of Russia; $55,000; principal investigator, Sergei Netesov; collaborator, Elizabeth Robertson, CDC; ISTC 883.

ℵ Monkeypox virus genome; $55,000; principal investigator, Sergei Shchelkunov; collaborators, Peter Jahrling, USAMRIID, and Joseph Esposito, CDC; ISTC 884.

ℵ Study of the genetic and serologic diversity of hantaviruses in the Asian part of Russia; $55,000; principal investigator, Lyudmilla Yashina; collaborators, Connie Schmaljohn, USAMRIID, and Stuart Nichol, CDC; ISTC 805.

ℵ Development of advanced diagnostic kit for opisthorchiasis in human patients; $55,000; principal investigator, Valery Loktev; collaborator, Victor Tsang, CDC; ISTC 691.

At the State Research Center for Applied Microbiology, Obolensk

ℵ Molecular-biological and immunochemical analysis of clinical strains of tuberculosis and mycobacteriosis; $138,000; principal investigator, Igor Shemyakin; collaborator, Thomas Shinnick, CDC; ISTC 810.

ℵ Investigation of the immunological effectiveness of delivery *in vivo* of the Brucella main outer membrane protein by the anthrax toxin components; $61,500; principal investigator, Anatoly Noskov; collaborators, John Collier, Harvard University, and Arthur Friedlander, USAMRIID; ISTC 919.

Projects being processed by ISTC as of October 1997

At the State Research Center for Virology and Biotechnology "Vector," Koltsovo

ℵ Experimental studies of antiviral activities of glycyrrhyzic acid derivatives against Marburg, Ebola, and human immunodeficiency virus; principal investigator, Andrei Pokrovsky; collaborator, John Huggins, USAMRIID.

At the State Research Center for Applied Microbiology, Obolensk:

ℵ Monitoring of Anthrax; principal investigator, Nikolai Staritsin; collaborator, Arthur Friedlander, USAMRIID.

The NAS committee used the following criteria in selecting the pilot projects:
Scientific importance of the topic: the project will make an important contribution to the epidemiology, prevention, diagnosis, or therapy of any disease that is associated with dangerous pathogen(s) or that is (1) historically linked with BW applications or (2) a source of substantial public health concern. If successful, the project will open up a new area of important research on dangerous pathogens.

---

2 Funds committed to U.S. collaborating institutions are: CDC, $47,000; USAMRIID, $20,000; Harvard University, $9,000.

Quality of the proposal: the project is scientifically and technically sound; anticipated results are clear; the project is feasible; there is a strong work plan; budget and manpower estimates are appropriate; and there are measurable milestones.

Quality or capacity of the principal investigator, research team, and facilities: the proposing laboratory must have strong technical capabilities in the general research area, and the personnel and facilities proposed must have adequate capabilities to carry out the project.

Provision for strong U.S. collaboration: the project involves a topic that will attract strong and relevant U.S. expertise, and the commitment of the U.S. collaborator(s) is clear.

Engagement of former Soviet BW expertise: the project involves former or current defense scientists, or facilities or it provides important contributions to a larger program that involves such scientists or facilities.

Promotion of transparency: the project meets standard ISTC access criteria, and reciprocal laboratory visits between collaborators are an integral aspect. Projects that meet such criteria and also offer access to facilities or personnel not previously engaged in collaborative projects are of particular interest.

The committee also made the judgment that each project's potential contributions to public health or U.S. national security interests outweigh the risk that the project might contribute to development or improvement of offensive BW capabilities.

The pilot projects were limited efforts, and the NAS committee concluded that the following additional criteria should be considered in the selection of projects within the larger program recommended in this report:

Likelihood of sustaining the research by attracting the interest of other organizations with financial capabilities to continue work in the general field after completion of the project; and

Promotion of linkages between Russian scientists working in institutions that had been involved in BW activities and those that were not involved in BW activities.

The committee also strongly recommends that, if future joint activities are pursued, U.S. specialists adopt a more proactive role in identifying possible research topics and proposals for funding. The framework for collaboration presented in this report is designed to promote this goal.

The NAS became a partner of the ISTC in Moscow, which provides an important administrative framework for processing and reviewing proposals, monitoring projects, and dispensing funds within Russia. Of special importance are the ISTC procedures for distributing funds for salaries directly to individual researchers, thereby circumventing opportunities for intermediaries to divert a portion of the funds for unintended uses.

Drawing on this first-hand experience, the NAS committee developed three overarching principles for guiding bilateral activities:

a.) *Projects should be collaborative in design and conduct*

Only projects that are of interest to specialists in both countries should be undertaken.

All projects should be conducted on the basis of cooperation and not assistance, with each side making intellectual, financial, and in-kind contributions.
All relevant constituencies in both countries should be able to apply for participation in the program.

b.) *Projects should be designed and conducted in a way that maximizes transpsrency*

Activities should be carried out in an environment of openness.
Direct contacts among specialists should be stressed.
A central coordination point within each government should be apprised of cooperative activities.

c.) *Results of cooperative projects should be disseminated to the widest possible interested audience*

Whenever possible research results should be promptly published or made available to international audiences through other channels.
Intellectual property and sensitive findings should be protected.
Intellectual property rights resulting from cooperative activities should be shared by the participating institutions on fair and equitable terms.

4. Recommendations for Future Activities

4.1 A PATHOGENS INITIATIVE

Although Russian interest in cooperation in this field is increasing, the future political course in Russia remains difficult to predict. As cooperation becomes more ingrained in the Russian scientific community, joint efforts are more likely to survive political shocks, thus underscoring the importance of establishing and broadening cooperation while the window of opportunity is open.
The core of a *Pathogens Initiative* should be joint research projects directed to the epidemiology, prophylaxis, diagnosis, and therapy of diseases associated with dangerous pathogens and related fundamental research. According to Russian colleagues, if the U.S. government decides to support such an initiative, early intergovernmental endorsement of the program could encourage the MOD to participate. In addition, such political support could help resolve many policy, implementation, and budget issues confronting Biopreparat and other interested organizations in both countries.

The committee recommends seven program areas as the initial framework for the program. The first five areas—anthrax, melioidosis/glanders, plague, orthopox virus, and viral hemorrhagic fevers—are agents/diseases that have been linked with BW activities for many years. In each of these areas the Soviet government is believed to have invested large financial resources in carrying out research that has been largely unknown outside the country. Organizations in the United States also have good research capabilities to help combat the infectious diseases of interest.

Two additional program areas will provide opportunities to address other pathogens/diseases of public health concern and to carry out related fundamental research. These two categories are particularly important in providing support for key Russian scientists who have become interested in pursuing careers not directly tied to potential BW agents and expanding the pool of potential collaborators in the United States.

A five-year program that builds to a level at which fifteen three-year projects are initiated each year, involving an average of ten full-time Russian specialists per project, could engage a substantial number of leading Russian specialists in the field and most of the key Russian research facilities. The projects will be selected on a competitive basis using the criteria set forth above. The resources devoted to each program area should depend on the quality of the project proposals across all areas. The pilot projects fall into several areas, and the possibility of expanding these limited efforts should be considered if the results are promising.

Several supporting activities could effectively complement the research activities. Specifically, upgrading the communications capabilities of selected Russian institutes, improving the safekeeping and utilization of strain collections used as national reference standards, and expanding exchanges of information on biosafety requirements and practices are appropriate areas of cooperation.

In addition to annual reviews of all projects, the overall approach will be evaluated in depth at the end of the second year and adjusted as necessary.

4.2 PHASE 2: AN ERA OF SUSTAINED COOPERATION FOLLOWING THE *PATHOGENS INITIATIVE*

Recognizing that sustained cooperation must be accompanied by rules of disclosure and other measures designed to provide assurance that work is devoted strictly to legitimate purposes, the NAS committee developed a model for progressive development of suitable transparency arrangements. The model is intended to emphasize the importance of transparency and to stimulate the official deliberations necessary to work out agreed-upon provisions. The model calls for an intergovernmental mechanism to provide direction for collaborative efforts on a broad front.

Joint research projects would continue to be the core of long-term cooperation. Expanded cooperation in epidemiology and rapid response to outbreaks of infectious disease would promote trust between the two countries. Related to the expanded international exchange of data, the internal capabilities of Russia to assess and process epidemiologic information would need to be strengthened. During outbreaks of diseases, specialists from the two countries should collaborate in providing their most relevant information and in offering technical support to each other. Also, Russian specialists who are involved in field investigations should be encouraged to apply for participation in the training programs of the CDC.

In addition, the promotion of effective national regulatory approaches to controlling dangerous pathogens appears to be essential to ensure responsible handling of dangerous pathogens on a broad front. Effective enforcement procedures and sharing of experiences are particularly important in developing mutual trust.

At the same time, the need to prevent dissemination of sensitive information to parties of proliferation concern and to protect intellectual property rights would continue to be important. The Russian and U.S. governments should adopt appropriate procedures to ensure that these issues are addressed in a manner that does not undercut the broader transparency objectives of cooperative endeavors.

5. Costs

To build on current momentum, the committee recommends that the U.S. Department of Defense (DOD) promptly provide financial support for the *Pathogens Initiative*. Sustained funding for the longer-term Phase 2 program would undoubtedly require agreement between the executive branch and the Congress about a line item in the budget of a selected agency. It is premature to speculate which department or agency should have long-term financial responsibility.

5.1 PATHOGENS INITIATIVE

The projected costs to the United States of the *Pathogens Initiative* are as follows: fiscal year (FY) 1998, $6 million; FY 1999, $7 million; FY 2000, $8.5 million; FY 2001, $8.5 million; and FY 2002, $8.5 million. Most of these funds are earmarked for direct project support.

5.2 PHASE 2: AN ERA OF LONGER-TERM SUSTAINED COOPERATION

As indicated above, it is assumed that the size of the program will grow steadily and then level off during the era of sustained cooperation. Under the model suggested as a goal for expanded efforts, the total U.S. contribution would be $10 million per year and the Russian institutions would be expected to cover most of their costs.

6. Anticipated Results of Program

The committee believes that the proposed joint efforts articulated in the NAS report could build a considerable level of trust between the scientific communities of Russia and the United States in a way that could help change the tone of diplomacy on the international security aspects of dangerous pathogens. Certainly, a likely effect of a multi-year program in Russia would be adjustment of the structure of the research enterprise dealing with dangerous pathogens. Research projects could be expected to be increasingly concentrated at a handful of the best institutions, which will become centers of excellence. To the extent that other institutions remain viable, they should be motivated to find work outside the area of dangerous pathogens. Thus, a continued, vigorous program of collaboration could also have profound direct and indirect effects in reducing the threats of proliferation and terrorism and improving U.S.-Russian relations in an area that too frequently has been punctuated with acrimony. Finally, the program would make many contributions to combating dangerous infectious diseases, while serving as a model for global efforts—a critically important goal, given the growing international recognition of the global dangers posed by new and reemerging infectious diseases.

Committee

Joshua Lederberg, *chair*, Professor Emeritus and Raymond and Beverly Sackler Foundation Scholar, Rockefeller University

John D. Steinbruner, *vice-chair*, Senior Fellow, Foreign Policy Studies Program, Brookings Institution

Barry Bloom, Investigator, Howard Hughes Medical Institute, Albert Einstein College of Medicine

Gail H. Cassell, Vice President of Infectious Diseases, Eli Lilly & Company

Robert Chanock, Medical Director, Laboratory of Infectious Diseases, National Institute of Allergy and Infectious Diseases, National Institutes of Health

R. John Collier, Presley Professor of Microbiology and Immunology, Department of Microbiology and Molecular Genetics, Harvard Medical School

Maurice R. Hilleman, Director, Institute of Therapeutic Research, Merck Research Laboratories

Peter Jahrling, Senior Advisor and Research Scientist, United States Army Medical Research Institute of Infectious Diseases

James LeDuc, Associate Director for Global Health, National Center for Infectious Diseases, Centers for Disease Control and Prevention

Matthew Meselson, Thomas Dudley Cabot Professor of Natural Sciences, Harvard University

Thomas Monath, Vice President, Research and Development, OraVax and Adjunct Professor, School of Public Health, Harvard University

Frederick A. Murphy, Professor, School of Veterinary Medicine, University of California-Davis

Philip K. Russell, Major General (retired US Army), Professor of International Health, School of Hygiene and Public Health, The Johns Hopkins University

Alexis Shelokov, Director, Medical Affairs, Biologics Development Center, The Salk Institute

Staff

Christopher P. Howson, Director, Board on International Health, Institute of Medicine

Jo Husbands, Director, Committee on International Security and Arms Control, National Academy of Sciences

Glenn Schweitzer, Director, Office for Central Europe and Eurasia, National Research Council

Chaarles G. Fogelgren, Research Assistant, Committee on International Security and Arms Control, National Academy of Sciences

La'Faye Lewis-Oliver, Administrative Assistant, Committee on International Security and Arms Control, National Academy of Sciences

Stephanie Y. Smith, Project Assistant, Board on International Health, Institute of Medicine

THE NEED FOR INTERNATIONAL COOPERATION TO PROVIDE TRANSPARENCY AND TO STRENGTHEN THE BTWC

Lev S. SANDAKHCHIEV
State Research Center *of* Virology and Biotechnology "VECTOR"
Koltsovo, 633159 Novosibirsk region, Russia

1. Introduction

Rapid advances in biology and biotechnology over the last few decades and even more promising results in the coming future increasingly attract the attention of the world scientific community and governmenal agencies in many countries.
The world expects significant benefits from advances in medicine, agriculture, environmental protection and many other areas of human life and activity.
In particular today's knowledge of the human genome has allowed for the development of new methods of genetic diagnostics and therapy. Significant funding is being spent on the creation of transgenic animals and plants. Genetic engineering applications are additionally becoming more and more widespread even in areas far distant from biology and medicine: industrial production, biorenewable energy resources, bioremediation etc.
Another expectation is that advances in biology will enable the world to withstand the growing threat of emerging and re-emerging infections. Mortality associated with infectious disease accounts for makes 33 percent of all deaths worldwide. New infections are appearing in significant numbers, while well-known illnesses thought to be under control are re-emerging.
According to the WHO data, 30 previously unknown infectious agents have been discovered over the past 20 years that cause such diseases as AIDS, hemorrhagic fevers, hantavirus pulmonary syndrom, etc. According to expert estimates, we currently know about only 30 percent of all existing malignant bacteria species and only 5 percent of viruses. These have stricken hundreds of millions of people worldwide. Especially toubling is the fact that no efficient preventive means against many of them are currently available.
Moreover, during recent years scientists have managed to discover the infectious nature of a number of already known diseases. The latest classical examples: viral hepatitis C and G, T-cell leukosis and peptic ulcer of stomach/duodenum (Helicobacter pylori).
And even those existing and known infections are changing. Unprecedented monkeypox and buffalo pox viruse outbreaks took place in Zaire and India in 1996-1997, taking numerous human victims. Unprecedented is also the ability of

E. Geissler et al. (eds.), Conversion of Former BTW Facilities, 149-156.
© 1998 *Kluwer Academic Publishers. Printed in the Netherlands.*

monkeypox virus to cause epidemics in humans. Moreover, it is currently feared that smallpox-related diseases may emerge and strike anew against humanity. With the eradication of smallpox in 1977 and cessation of vaccination against this disease, the danger has become greater for nonvaccinated people to become infected with monkeypox or cowpox (buffalo pox).

2. Emerging and re-emerging diseases

In this way, the formerly existing stable view of infectious agents has been replaced by a regard for their dynamically changing properties in antigenic structure, drug resistance, and many others.

2.1 VIRAL INFECTIONS

Arboviral infections, borne by various insects, still remain some of the most critical infections being presented worldwide. Just a simple citation of these infections demonstrates how important this issue is. They include such severe viral diseases as Japanese encephalitis, tick-borne encephalitis, dengue fever, Omsk hernorrhagic fever, yellow fever, and many others, some of which are currently considered as "aggressive infections". Thus, despite the existence and use of many prophylactic and therapeutic drugs, the incidence of tick-borne encephalitis in Russia increased by 61 percent reaching 10 thousand cases, last year. And, once having come from Japan, this infection has spread out, reaching Europe. A similar pattern is observed with other viruses. A survey of the Venezuelan equine encephalomyelitis (VEE) outbreak in South America revealed a new pathogenic form of human virus variants that caused disease in thousands of people. Urban outbreaks of yellow fever have demonstrated how arboviral infections can evolve and adapt to changing environmental conditions in highly populated areas, and epidemic outbreaks of dengue fever strike millions of people in different countries worldwide. All this make it absolutely necessary to enhance research in this area.

Viruses causing hepatitis E and C discovered in 1988 and 1989 belong to the emerging viral infections as well. These diseases are widely spread in Russia, and in some regions up to 30 percent of infectious hepatitis cases are caused by these very viruses.

Pathogens of hemorrhagic fevers are of special importance. For the last few decades a number of viruses have been identified that cause such diseases as Ebola hemorrhagic fever (Ebola virus, 1977), hemorrhagic fever with renal syndrome (Hantaan virus, 1977), hantavirus pulmonary syndrome (Sin Nombre, 1993), and Brazilian hemorrhagic fever (Sabia virus, 1994). Outbreaks of hemorrhagic fever with renal syndrome occur periodically in Russia. Thus, the 1997 outbreak has stricken 4,300 people.

2.2 PARASITIC INFECTIONS

Many parasitic infectious that still remain widely spread in Russia and other countries worldwide also belong to the emerging infections. The most common parasitic infection in Russia is *opistorchiasis,* which is associated with river fish consumption. The highest incidence rates are in Siberian regions with lots of natural foci of the disease. According to Russia's official statistics, over 30,000 cases of opistorchiasis are registered annually, although these data are not accurate due to insufficient diagnostics; excerpted data of epidemiological surveys indicate a significantly higher incidence. Epidemiologists expect several million inhabitants of the endemic regions in the east of the country to be infected. Other peculiarities of parasitic infections are connected with their changing clinical manifestation and growing risk groups. The fact that more and more people appear to have impaired immunity, e.g. AIDS-patients, results in dramatically increasing mortality of such patients from *toxoplasmosis.* Extensive migration may lead to new species of parasites being brought into regions where they were not known before and whose diagnosis is often impossible due to the lack of specialists and necessary test kits.

2.3 BACTERIAL INFECTIONS

One third of the population worldwide (2 billion people) is infected with *Mycobacterium tuberculosis,* the pathogen causing TB. Each year this tremendous figure increases by 8 million new cases. For the coming decade, the estimated increase in TB incidence will be 50 percent in South-Eastern Asia, and 100 percent in Africa. The increasing rates of TB cases are considered to be associated, among other reasons, with the spread of HIV infection. Besides HIV infection, another important factor contributing to TB dissemination is the emergence of *M. tuberculosis* strains resistant to antibiotics.

TB may be caused in humans not only by *M. tuberculosis* but also by other components of the so-called TB-associated complex: *M. bovis* and M. *africanum. Mycobacterium bovis,* responsible for bovine TB, has many other animal hosts. Currently, TB is estimated to be the cause of 7 percent of deaths and 27 percent of avoidable deaths. The epidemiological situation about TB in Russia has been deteriorating since 1992. The highest rates are observed in Western Siberia and the Far East.

3. The State Research Center of Virology and Biotechnology "VECTOR"

"VECTOR" is a large research and production complex whose primary activities are focused on basic and applied research in the areas of theoretic virology, molecular biology, virology, immunology. aerobiology, epidemiology, biotechnology as well as development and manufacture of preventive, therapeutic and diagnostic preparations.

Though it has never been directly involved in BW design or production, before 1989 it was primarily oriented to defensive programs research and up to 90% of its research funding came from those programs. It may be considered a former defensive BW facility according to the terms discussed at the NATO ARW in Budapest (see Annex 1: Glossary).

During the period 1974-1991, the Center created an advanced scientific base allowing research of highly dangerous infectious agents at Biological Level 3 (BL-3) and BL-4: theoretical and experimental studies in the area of genetic engineering, biotechnology and aerobiology.

The staff of "VECTOR" totals some 1,200 researchers, including 4 academicians, 17 Doctors of Science and 166 PhDs. The research and technical staff is represented by highly qualified personnel specialized in the fields of genetic engineering, molecular biology, virology, theoretical virology, immunology, epidemiology and ecology, with significant experience with viral pathogens and the production of diagnostic and prophylactic preparations.

The total area of our research and production facilities amounts to more than 250 thousand square meters; our property covers 8,000 hectares (19,768 acres). Our main research and production area located at Koltsovo, Novosibirsk region, having some 14,000 residents. Our facilities have all the necessary infrastructure, including transportation, electricity, water and energy, sewage treatment, etc.

Several buildings intended for work with highly dangerous infections meet the specific safety demands: an air-tight external envelope, negative pressure in the working area, complete sterilization of all types of discharge (gas, liquid, and solid), etc. The WHO commission in 1995 confirmed that the working conditions in the Center meet the international biosafety requirements.

SRCVB "VECTOR" is one of the two scientific and experimental complexes in Russia that mainly focus on virus infections research. The State Research Center for Applied Microbiology (Obolensk, Moscow region) is a similar scientific and experimental facility involved in bacterial infections research. SRCVB "VECTOR" and SRC for Applied Microbiology are the only bases in Russia and the rest of the NIS countries that allow studies of highly dangerous pathogens.

Our researchers work in close collaboration with the scientists of the Siberian Branch of the Russian Academy of Sciences and the Siberian Branch of the Russian Academy of Medical Sciences, also located in Novosibirsk.

The collection of cultures of microorganisms available at the Center comprises over 10,000 entries: various viral strains including the national collection of variola virus strains, isolates of viral strains, recombinant viral strains, strains of microorganisms including producer strains, etc. The collection of cultures of microorganisms has received international recognition: The Collection of the SRC VB "VECTOR" was affiliated with the European Culture Collection Organization (ECCO).

"VECTOR" is one of the two WHO Collaborating Centers in the world for orthopoxvirus diagnosis and a repository for variola virus strains and DNA, with one of its goals being to preserve and study the Russian collection of various virus isolates.

The Center has a breeding farm for laboratory animals, primates included, for performance of various trials. The monkey breeding facility is presently the only one active in Russia. The facilities allow preclinical and clinical trials of new medicinal preparations.

3.1 "VECTOR" R&D RESTRUCTURING

After the dramatic decrease in defense-related projects in 1991 and general cuttings of the state funding due to the economic situation in Russia, the Center had to reconsider and redirect its R&D activity on the fundamental and applied problems of public health, veterinary medicine, and environmental protection, as well as to establish production of medical and diagnostic products for public health and veterinary medicine. Federal funding has been radically re-directed from defense-related projects towards fundamental and applied research for human application. The largest part of our funding (89%) comes to the State Research Center of the Russian Ministry of Sience and Technologies. The funding from other sources is relatively small.
The main goals of "VECTOR" R&D restructuring are listed below:
* Reorient our personnel having extensive experience in the areas of virology, genetic engineering and biotechnology to carry out research for the needs of public health, agriculture and environmental protection.
* Carry on basic research in the area of molecular virologv and virus mutability.
* Enhance applied research aimed at development of new therapeutic, preventive, and diagnostic agents for public health and veterinary medicine employing the latest achievements in genetic engineering and biotechnology.
*Develop production facilities to manufacture a wide range of drugs for public health and veterinary medicine.
*Collaborate internationally both in basic and applied research, and drug production.
To achieve the goals we carried out structural reorganization of the Center by assigning research institutes to specific areas of basic and applied studies, thereby establishing production and marketing facilities, and adapting to these servicing and supporting structures.

3.2 "VECTOR" PRODUCTION FACILITIES

During the last few years the Center has managed to create its own production facilities to produce different diagnostics kits, vaccines, immunobiological preparations, laboratory-use products, etc. numbering more than 150 items. Most of these preparations were designed at our Center.
The reconstruction of our facilities has enabled "VECTOR" to establish several advanced production sites for cultivation, purification, and manufacturing of dry and liquid injectable drugs, as well as prepackaging sites (blister and vial packaging) for ready-to-use drugs.
In addition, that we have been efficiently cooperating with the following foreign pharmaceutical companies to manufacture drugs from substances bring imported to Russia: APOTEX (Canada), Bayer AG (Germany), Norton Healthcare (Great Britain). KRKA (Slovenia), LEK (Slovenia), POLFA (Poland). To be exact, this cooperation is currently limited to final stages of drug production: pre-packing and quality control. Even so, this has become a very profitable business in that it makes it possible to identify and promote the most promising drugs to the Russian market without much expense.

In order to improve further the production facilities we'll have to undertake the following:

* obtain international GMP certification:
* expand international cooperation with pharmaceutical companies;
* facilitate innovative projects to produce new vaccines and treatments that are in demand in domestic and international markets. The Center is one of the top ten most efficient and dynamically developing pharmaceutical manufacturers in Russia in terms of sales volume and list of products. We operate throughout Russia through an efficient network of distributors. Thus, our production and sales are making "VECTOR" less dependent on state funding. In 1996 the income from sales of products and sevices reached 65% while it was only 43% in 1990. Nevertheless we consider international grants (ISTC, CRDF, INTAS and some others) to be a significant source of support for our research funding and of great value for the stable relationship with the world scientific community. More than thirty of our scientists are currently working in several research institutions abroad, which gives us an excellent opportunity to establish closer contacts with these facilities.

In the area of virology and molecular biology, we actively collaborate with 12 leading research centers abroad, including the Center for Disease Control and Prevention in Atlanta, the National Institute for Allergy and Infectious Diseases, Bethesda, etc. A Program for Research and Monitoring of Pathogens of Global Importance was initiated this year.

Currently, we have submitted over 60 project proposals for review by the US Department of Energy under the Industrial Partnering Program, which is probably the only program that anticipates of mutually beneficial commercial activity. The IPP Program is in many ways simply following the global trend for establishing strategic business alliances at an international level.

We also actively collaborate with the US National Academy of Sciences in the control of pathogens of global importence and participate in the Program for Countering Emerging Infectious Diseases (ProCEID).

"VECTOR" is interested as well in participating with other than the above mentioned international programs: WHO programs on developing of new vaccines, NATO Scientific Committee programs, ProCEID and others and we are prepared to discuss with any potential partners additional projects that may be considered as proposals for bilateral or international cooperation or projects ready for direct commercialization.

We may conclude that insufficient infection control and surveillance, the broadening migration of populations, the growing number of people with impaired immunity, the changing environment and the wide range of experiments in biotechnology and genetic engineering, along with many other factors, will contribute to the growing world concern over the potential threat of the infections.

It is most vital, therefore, to establish an **International Laboratory for Emerging Infections** whose goals would be to investigate infectious diseases so as to predict and prevent their spread, and develop diagnostic, preventive and therapeutic drugs with appropriate preclinical and clinical trials in full compliance with the international requirements. It is proposed to utilize the experimental and clinical base of the *International Laboratory* for studies to be performed by scientists from other research institutions of Russia and those of the state-parties to this project (20 to 30 scientists a year).

The Department of Basic Medicine of the Novosibirsk State University is based in the SRC VB, thereby allowing us to engage students and graduates of the University in the work of the *International Laboratory*.

Thus, taking into consideration the need to enhance research and development in order to prevent infectious disease the need to internationally unite efforts to withstand the growing threat of emerging infections and the possibility of biological terrorism and the best possible opportunity to secure the confidence of the parties involved on the issues of BWC verification, we propose to establish the *International Laboratory for Emerging Infections* at the State Research Center of Virology and Biotechtiology "VECTOR" of the Russian Ministry of Public Health.

This Laboratory could operate under the aegis of the WHO and within the framework of existing or future international programs aimed to prevent infectious disease and biological terrorism. The activity of this Laboratory might be financed by grants from international organizations, government agencies and free market investors, or an International Fund might well be organized.

The proposed main goals of the International Laboratory:

1. Development of a system for monitoring of "emerging" and "reemerging" infectious diseases;
2. Isolation and analysis of presently circulating infectious pathogens: evaluation of their natural evolution, and investigation of their molecular epidemiology;
3. Organization and perfomance of field expeditions to the territories of natural foci and to the regions with suspected occurrence of pathogens of "threatening" and emerging infections;
4. Diagnosis and confirmation of complicated infection cases;
5. Development of new molecular techniques for diagnosis and characterization of "emerging" infections (basing on microchips and phage display);
6. Theoretical analysis of the primary structure of the pathogen isolates obtained, and creation of a database for the primary structures of "emerging" infectious pathogens;
7. Development of recommendations on prevention of infectious outbreaks and measures for combatting infectious diseases.

To ensure the efficient operation of this International Laboratory it should have access to the latest information technologies including telemedicine that along with the INTERNET will enable medical teleconferencing concilia, precise diagnosis and proper symptomatic and specific treatment.

4. Strengthening the BWC by Transparency

A few comments regarding the issues of transparency and BWC strengthening.

First of all, there should be no doubt about the determination of the new Russia to promote international cooperation, especially cooperation aimed at the prevention of proliferation of biological and toxin weapons.

In April 1992 the Russian President signed Decree #390 prohibiting the conduct of any research that violates the UN Biological Weapons Convention on the territory of Russia. These regulations were introduced into the criminal code of Russia. Import/export operations of pathogenic microorganisms are also subject to licensing. A law has also been passed that provides for government regulation of any genetic engineering research.

Annually our Center reports to the UN Organization on the use of BL-3 and BL-4 laboratories involving research on highly dangerous pathogens. The work in this area is also subject to control by the Russian Committee on Issues Relating to the BW Convention in the Office of the Russian President. All this demonstrates that the political and legal environment in Russia is completely consistent with the proliferation prevention concept. Beyond this, however, our main efforts must be directed towards involvement of specialists and production capabilities for civil application and prevention of the outflow of researchers and technologies to places with an unstable political environment.

Viral and bacterial injections control systerns cannot be efficiently implemented locally in just one country. It is necessary to establish the exchange of information and viral strains that have not yet lost their natural properties and are not yet resistant to therapeutic drugs and antibiotics.

Taking into account the need to oppose the growing global threat of emerging and re-emerging disease it is vital not to destroy existing facilities for research of highly dangerous pathogens, but rather to preserve them and to use them cooperatively to counter infections. In this respect the idea of the **International Laboratory** is in full compliance with Article X of the Convention and the objectives of ProCEID. This may become a model of international cooperation, using advances in biology and biotechnology not for destructive purposes but for the benefit of human health. Once the idea becomes real and productive, a number of such International Laboratories might be set up in other countries posessing similar facilities.

Acknowledgments

I gratefully acknowledge Mrs. Raisa Martyniuk, Mr. Evgeny Starkov, and Mr. Vladimir Ryabenko for their careful preparation of the manuscript and for technical assistance.

REDIRECTION OF BW EXPERTS IN THE FRAMEWORK OF THE INTERNATIONAL SCIENCE AND TECHNOLOGY CENTER (ISTC)

JOHN COMPTON AND DIANA POBEDIMSKAYA
International Science and Technology Center
Luganskaya ulitsa, 9
Moscow, Russian Federation

1. Introduction

The founding of the ISTC was a direct result of the dissolution of the Soviet Union and resulting end of the Cold War. The former Soviet defense sector was highly destabilized and a host of science and technology institutes faced drastic reductions in operating funds. As a consequence, scientists who had formerly enjoyed a privileged status were faced with marginal salaries and few prospects for a meaningful career, either in their old defense-related activities or in alternative conversion activities. The rest of the world then faced the prospect of the proliferation of the expertise and knowledge of a generation of former Soviet weapons scientists whose work and lives had been severely disrupted. The main concerns were for scientists involved with weapons of mass destruction, namely, nuclear, chemical, biological, and delivery systems.

The ISTC was invented as a mechanism which could give weapons experts in the Commonwealth of Independent States (CIS) the opportunity to redirect their talents to peaceful activities and contribute to solution of national and international science and technology problems. The founders further hoped to reinforce the transition to market economies, support basic and applied research, and promote integration of CIS scientists into global scientific community.

2. Brief History and Status of the ISTC

The ISTC is an intergovernmental organization founded by the European Union (EU), Japan, Russia and the USA in November 1992, and operations began in March 1994. It is registered as a diplomatic mission with the Russian Ministry of Foreign Affairs, and as such it is free of taxes and customs duties. Of the former Soviet republics Armenia, Belarus, Georgia, Kazakhstan and the Kyrgyz Republic have now joined the organization. Norway acceded to the ISTC in 1997 and the Republic of Korea has formally initiated accession procedures as well.

The ISTC has a Governing Board, its decision-making body with four permanent and one rotating seat. There is also an independent Scientific Advisory Committee which

E. Geissler et al. (eds.), Conversion of Former BTW Facilities, 157-165.
© 1998 *Kluwer Academic Publishers. Printed in the Netherlands.*

meets regularly. Approximately one hundred people work at the Secretariat, the executive body located in Moscow. The premises are provided by the Russian Federation, and the staff is detailed by the four initial Parties.

The ISTC funds a broad range of projects in support to non-proliferation regimes. Some examples include environment assessment and monitoring, nuclear power plant safety, nuclear waste disposal, basic physics, lasers, space science, materials science, instrumentation, medical equipment, and new drugs and vaccines, but this is hardly an exhaustive list (see Table 1). As of 15 November 1997, 1404 project proposals had been registered (1168 from Russia and 236 from other CIS states), 496 projects approved and funded, and $154 million (USD) committed. Of these funds, the US provided $66 million, the EU $60 million, Japan $25 million, and other sources $3 million. Over 280 institutions and more than 17,000 specialists currently receive grants from the ISTC.

TABLE 1. ISTC-funded projects by technology area

Technology Area	Fraction of Total Support
Environment	25%
Fission Reactors	19%
Physics	12%
Materials	11%
Biotechnology and Life Sciences	8%
Instrumentation	7%
Fusion	5%
Information and Communication	4%
Space, Aircraft and Transportation	4%
Chemistry	2%
Non-nuclear Energy	2%
Manufacturing Technology	1%

3. Advantages of the ISTC

The ISTC has a number of distinct advantages over other programs organizations that fund science projects in the CIS. As was mentioned earlier, these derive from the governmental recognition of diplomatic mission status of the organization. First, it has an in-country project management infrastructure and provides both technical and financial on-site monitoring of projects. Furthermore, the financial operation is noteworthy in that scientists are paid directly through fund transfers that are not subject to taxation. Similarly, procurements are handled expeditiously because of tax and customs exemptions. Other advantages include fair and reasonable terms for intellectual property rights (IPR), and training in IPR issues is offered by the Center in various locations as a means of orienting scientists towards integration into the global economy. Finally, the Center protects proprietary information of both funded projects and sponsoring organizations.

In June 1997 the ISTC formally initiated its new Partners Program as a way to attract alternative support for projects, namely through government agencies, other intergovernmental organizations, non-governmental organizations, and the private sector. It furthers the aim of ISTC to help projects become eventually self-supporting by contacts with foreign industry and other organizations. Some organizations that have joined in the program are CERN, Swiss NC-Laboratory, US National Academy of Sciences, Framatome, Dow Chemicals, 3M, Scientific Utilization Inc., Rhône-Poulenc, Dupont, Sandia National Laboratories, the US Department of Energy, and Burlington Biomedical and Scientific Corp. The Partners Program can save sponsoring organizations at least half of their expenditures through tax and customs exemptions and also through direct payment to scientists. (Institutes do, however, receive a small fraction of project funds as overhead.) The biotechnology area in particular has benefited from the new program.

After nearly four years of operation, the ISTC has established itself as a mature intergovernmental organization. It is now able to devote more resources to the utilization of project results and technology implementation. By continuing to expand its public outreach efforts it can become more widely known throughout the world and serve as a clearinghouse of information and guide for those seeking cooperation with CIS scientists. This in turn leads to improved links with other international organizations and helps to attract new sources of funding.

4. ISTC Seminars Program

Over the past three year the ISTC has sponsored eleven seminars, five of which were devoted to biotechnology and life sciences. A summary of these seminars offers an excellent overview as to its involvement in BW conversion efforts. A main objective of the seminar program is to stimulate the development of new proposals in cooperation with foreign scientists by providing opportunities for contacts between Western, Japanese, and CIS scientists. This facilitates mutual understanding and helps identify mutual interests in a wide range of research fields as a first step towards collaboration. It also improves proposals and projects by stimulating a synergistic effect among these scientists. In particular it defines promising areas where collaboration would be most beneficial and productive in a specific field through intensive discussion between the foreign and CIS scientists on the current status of research. Thus seminars promote the further integration of CIS scientists formerly involved with weapons of mass destruction into the international scientific community. The following list provides the essential details concerning each of the five seminars:

(1) Title: Symposium on Development of Vaccines for Viral Infections
Date: December 12-14, 1994
Location: Russia, Koltsovo (Novosibirsk), "Vector"
Participants: 29 in total: 16 from Russia, 8 from the EU, 3 from the USA, 2 from the ISTC Secretariat.

(2) Title: Development of Vaccines against Bacterial Infections, Pre-clinical and Clinical Trials
Date: April 3-6, 1995
Location: Russia, Obolensk, Applied Microbiology
Participants: 34 in total: 19 from Russia, 8 from the EU, 4 from the US, 3 from the ISTC Secretariat.

(3) Title:	Development of Vaccines Against Severe Infections; Ethical Aspects of Clinical and Pre-clinical Trials
Date:	December 11-16, 1995
Location:	Russia, Pokrov, Pokrov Biologics Plant
Participants:	About 90 in total: 7 from the EU, the same number from the USA, 2 from Ukraine, 2 from Belarus, 2 from Kazakstan, 68 from Russia as well as 5 from the Secretariat and 4 Governing Board representatives.

(4) Title:	The Development of Medical Equipment Technology Converted from Weapons Scientists
Date:	February 19-23, 1996
Locations:	Russia, Podolsk, Moscow, Lutch Region
Participants:	About 100 in total: 73 from Russia, 11 from the EU, 3 from the US, 1 from Japan, 9 from the Secretariat as well as the Representative of the Administration of the President of the RF, Minatom, the Ministry of Science and the Ministry of Health.

(5) Title:	Severe Infectious Diseases: Epidemiology, Express-Diagnostics, and Prevention
Date:	June 16-20, 1997
Location:	Russia, Nizhne-Ivkino, Volgo-Vyatski State Scientific Center of Kirov Region Applied Biotechnology
Participants:	There were 25 participants from the CIS, 7 from Japan, 12 from the US, and 3 from the EU.

5. Statistics on ISTC Projects in Biotechnology and Life Sciences

As of December 1997, the ISTC had received 113 proposals at various stages of development related to biotechnology and life sciences. More than 50 Russian and CIS institutions are involved in these projects as leading institutes. The following institutions have been the most active in submitting projects in the field of advanced biotechnology and life sciences:

1. State Research Center of Virology and Biotechnology "Vector", Koltsovo	16
2. State Research Center for Applied Microbiology, Obolensk	13
3. Institute of Engineering Immunology "Biopreparat", Lyubuchany	8
4. GosNIIOKhT, Moscow	6

The total number of projects funded is 43. The allocated amount of funds is $10,056,121 or 8% of total ISTC funds. The involvement of ISTC Parties in the funding of the projects is shown in Table 2.

TABLE 2. Number of projects funded in biotechnology and life sciences

Funding Party/ Co-funding Party	EU	Japan	US	Others	Total
EU	2	4	23	3	32
Japan	4	-	3	-	7
US	23	3	17	3	46
Others	3	-	3	1	7

Project Agreements have been signed for 25 projects. The other 18 projects are waiting for signature of a Project Agreement. The total budget is $5,332,149. Table 3 gives the distribution of support according to ISTC's internal technology codes.

TABLE 3. ISTC-Funded projects by technology code
(Biotechnology and life science only)

Technology Code	Fraction of Total Support
Biochemistry	2%
Cytology, Genetics, Mol. Biology	19%
Ecology	4%
Immunology	12%
Microbiology	8%
Pharmacology	26%
Public Health	20%
Radiology	7%
Nutrition	2%

Biochemistry includes studies of chemical processes in biological systems and identification and measurement of biochemical substances. The number of ISTC's projects categorized in this group is small. A typical example of this group is #866, which proposes the study of the molecular mechanisms of the generation of electrical currents by bacteriorodopsin, a membrane protein. As a result of the project's implementation, mechanisms will be clarified for key pigment-protein complexes responsible for transformation of light energy into electric energy within the membranes of photosynthetic organisms. The leading institution of #866 is Moscow State University.

Cytology, Genetics and Molecular Biology includes studies of genetic apparatus, structure and functions of proteins and nucleic acids, and mechanisms and functions of living cells and cells components. This group is composed of:

(1) Study of genetic and serologic diversity of different epidemic viruses:

#622 Conservation of the National Virus Collection, D.I. Ivanovsky Institute of Virology

#753 Genetic elements of viable viruses in the permafrost region of Russia, State Research Center of Virology and Biotechnology "Vector"

#805 Genetic and serologic diversity of hantaviruses in the Asian part of Russia, State Research Center of Virology and Biotechnology "Vector"

#883 Molecular variability of hepatitis C virus in the West Siberian part of Russia, State Research Center of Virology and Biotechnology "Vector"

#884 Monkey pox virus, State Research Center of Virology and Biotechnology "Vector"

(2) Recombinant vaccine technology utilizing the bank culture discovered in Russia:
#120-2 Recombinant vaccines against diphtheria, State Research Center for Applied Microbiology
#131 Recombinant vaccines against hepatitis B, State Research Center of Virology and Biotechnology "Vector"
#132 Recombinant vaccines against hepatitis A, State Research Center of Virology and Biotechnology "Vector"
#337 Recombinant vaccine against legionellosis, State Research Center for Applied Microbiology

(3) Research on molecular or cellular biology:
#055 Research of hypervariable genome sequences, Institute of Gene Biology
#889 Phospholipid vesicles' application for gene therapy, GosNIIOKhT
#890 Novel biosensors based on bacterial cells, State Research Center for Applied Microbiology

Ecology includes effects of polluted environments on organisms and on water and terrestrial ecology. ISTC's projects related to this group aim to develop a new technology for intensive treatment of polluted soil and water based on natural microorganisms-distracters.

#088 Regional raw materials utilization, NPO "Biochimmash"
#119 Technology for the elimination of environmental oil pollution, State Research Center for Applied Microbiology
#228 Biological treatment of polyphenyl polluted soils, Research Center of Toxicology and Hygienic Reglamentation of Biopreparations
#440 Biological recultivation of soils polluted with radionuclides, Research Center of Toxicology and Hygienic Reglamentation of Biopreparations

Immunology includes mechanisms of immune responses, structure and functions of antigens and antibodies, as well as immunopathology and immunochemistry. Within this group the ISTC's projects can be divided into three main directions:

(1) Development of gene engineering immunomodulating preparations: new cytokines, recombinant immunotoxins with various structures and different modes of action for the treatment of common cancers and new recombinant protein-peptide immunomodulators:
#091 Design of new recombinant protein-peptide immunomodulators, Institute of Immunological Engineering of Russian Stock Company "Biopreparation"
#402 Recombinant immunotoxins against prostate cancer cells, State Research Center for Applied Microbiology
#693 Study of the structure and functions of the new cytokine, Institute of Immunological Engineering of Russian Stock Company "Biopreparation"

(2) Elaboration of new enzyme immunoassay kits for clinical detection of a number of infectious diseases, such as diphtheria, tuberculosis, mycobacteriosis, toxoplasmosis:

#689 Enzyme immunoassay kit for diagnosis of toxoplasmosis, State Research Center of Virology and Biotechnology "Vector"
#691 A new enzyme immunoassay for clinical detection of human liver fluke opisthorchis, State Research Center of Virology and Biotechnology "Vector"

(3) Development of live vaccine preparations:
#940 A new active culture influenza vaccines, State Research Center of Virology and Biotechnology "Vector"

Microbiology includes research on new ecologically safe biopesticides and microorganisms for the degradation of synthetic polymers and toxic chemicals.

#363 Development of a new generation biopesticides, State Research Center for Applied Microbiology
#593 Development of environmentally safe bacterial biopesticides, GP Biotechnological Innovation Center
#667 Highly efficient rhizobial strains to increase the yields of legumes, Research Center of Toxicology and Hygienic Reglamentation of Biopreparations
#A-092 Database of microorganisms for the degrading synthetic polymers and toxic chemicals, Institute of Microbiology, Yerevan

Pharmacology is the largest group, about 30% of all projects. It includes synthesis, composition, properties and effects of drugs. The main directions of this group among ISTC projects are the following:

(1) Research on drugs with anti-HIV and antiviral activity:
#127 New nucleoside conjugates with phospholipids, GosNIIOKhT
#319 Nucleotide inhibitors without intercell phosphorylation, GosNIIOKhT
#633 Organogermanium compounds, GNIIKhTEOS
#801 Additional application of MYRAMISTINE, NPO "Biochimmash"

(2) Research on antitumor drugs:
#158 O18 heavy water for a tumor therapy, NIIIT (Pulse Technique) and Cancer Research Center
#464 New antitumor ferrocene-based compounds, Ultrasan Corp.
#669 Rubomycine extracted from ray fungus culture, JSC Omutninskyi Scientific, Experimental and Industrial Base

(3) Research on cardiac drugs:
#123 Synthesis and study on nitrate compounds, Institute of Chemical Physics
#163 Search for Ca-channel blocking compounds, GosNIIOKhT

(4) Others:

#133 Live measles vaccine for oral administration using Russian's L-68 culture, State Research Center of Virology and Biotechnology "Vector"

#341Ecologically pure antibiotics production - direct sorption method, A.A. Bochvar All-Russian Scientific Research Institute of Inorganic Materials

#403 Research on *de novo* proteins, Institute of Immunological Engineering

#420 Development of new drug delivery systems, State Research Institute of Highly Pure Biopreparations

#421Recombinant protein technologies, such as superoxide dismutase, State Research Institute of Highly Pure Biopreparations

Public Health includes protection and improvement of community health, toxic and infectious disease control, preventive medicine and also detection equipment. This group includes ISTC projects concerning (1) the development of medical technologies for early diagnostics of malignancies, infectious diseases, etc.; and (2) monitoring of dangerous infections and creation of a data base.

(1) Medical technologies:

#280 Optical 3-D-Intrascopy-Medical Optical Tomography, Vavilov State Optical Institute

#475 Laser equipment for tumor diagnostics, NIKIET

#506 Express diagnostics of diphtheria, Research Center of Toxicology and Hygienic Reglamentation of Biopreparations

#621 Luminescent immunoassay diagnostics for screening newborns, State Research Institute of Biological Instrument Making

(2) Monitoring of dangerous infections and creation of a data base:

#550 Epidemiologically dangerous viruses with genome variability, D.I. Ivanovsky Institute of Virology

#637 Spatial trends of anthrax pathogen contamination in Russia, Central Research Institute of Epidemiology

#K-159 Monitoring of dangerous infections in Kazakstan, Institute of Informative and Control Problems

Radiobiology includes the development of radio-medical technology and the study on biological effects of radiation doses.

(1) Development of radio-medical technology:

#326 X-ray irradiator on the basis of the miniature pulsed accelerator ARSA, VNIIEF

#392 Highly sensitive X-ray luminescent ceramics screens which enables to decrease radiation exposure dose for patients, Vavilov State Optical Institute

#828 New process for production of medical isotopes from the liquid fuel reactor, which is more productive with less radioactive waste, FEI (IPPE)

(2) Study on biological effects of radiation doses:

#770 Tritium and tritium oxide effect on human organism, VNIIEF

#773 Reconstruction of absorbed doses by the rate of stable and unstable chromosome aberrations in nuclear specialists, Institute of General Genetics
#947 Methods of minimizing the medical consequences of radiation exposure, Branch-1 State Scientific Center RF - Biophysics Institute

Nutrition includes development of feed and food additives for humans and animals.

#802 Acidified milk preparations by spray drying method, NPO "Biochimmash"
#803 Food additives on the basis of Spirulina microalgae biomass, NPO "Biochimmash"

6. Conclusions

Projects in biotechnology and the life sciences are playing an increasingly important role in the overall ISTC activities. Out of the most recently funded group of projects over half were in this area, a rather significant increase. Non-traditional funding is certainly an important factor in this trend. For example, the US National Academy of Sciences funded six projects in the past year as part of the Partners Program. Since the last half of 1997 represents only the beginning of the Partners Program it can be expected that an increasing fraction of projects funded in the biology area will follow a similar pattern. The ISTC is actively attempting to expand the scope of CIS institutes and scientists that participate in its activities.

CONVERSION OF BW FACILITIES IN KAZAKSTAN

ANTHONY RIMMINGTON
Centre for Russian & East European Studies
The University of Birmingham
Edgbaston
Birmingham B15 2TT

Recent analysis of the issues arising from the dismantlement of the Soviet offensive biological weapons programme has focussed on the Russian Federation [1]. Given that Russia has inherited most of the organisational and physical infrastructure associated with Soviet development and production of biological weapons then this is perhaps not surprising. However, this paper will seek to demonstrate that among the newly independent states Kazakstan must be considered to be of especial importance with regard to the conversion of biological weapons since certain key elements of the Soviet offensive BW programme were located on, or contiguous to, its territory. Evidence will be presented which suggests that in terms of BW capability within the Former Soviet Union Kazakstan's biopharmaceutical industry is second only to that of the Russian Federation itself.

The present predicament of Kazakstan's BW facilities has much in common with that experienced by their Russian counterparts and gives rise to many of the same concerns amongst the international community. Has capacity for the development and production of biological weapons been dismantled and have military microbiological facilities been brought fully under the control of civil bodies? Can an outflow (the so-called brain drain) of military microbiologists to countries intent on pursuing their own BW programmes be prevented via a combination of international and bilateral initiatives? Can adequate safeguards be put in place to maintain the security of Kazakstan's BW facilities and their collections of pathogenic microorganisms and toxins? However, certain aspects of the situation with regard to BW facilities in Kazakstan are unique and pose particular problems for agencies concerned with arms control and verification. Chief among these is the fact that Russian personnel who in the recent past, have had extremely close links to the Soviet/Russian military-industrial complex, continue to occupy key management and scientific positions both in individual BW facilities and in the Kazak biopharmaceutical industry as a whole.

This article seeks to provide an overview of Kazakstan's military microbiological facilities and the role they were allocated within Soviet BW programmes administered by the USSR Ministry of Defence, *Biopreparat* and other organisations. It argues that the Kazak authorities have demonstrated a strong commitment to the conversion process via a number of initiatives. These include attempts to institutionalise the

E. Geissler et al. (eds.), Conversion of Former BTW Facilities, 167-186.
© 1998 *Kluwer Academic Publishers. Printed in the Netherlands.*

conversion process via the creation of new civil organisations such as the National Centre for Biotechnology which incorporates virtually all facilities in Kazakstan which were engaged in BW R & D and production. The government has also sought to neutralise military activity at the Stepnogorsk complex, which represents the most substantial BW capability in Kazakstan by selecting it as the hub for the development of a new modern civil biopharmaceutical industry. However, despite this evident commitment to demilitarisation the authorities have not released adequate resource. It is argued that for the conversion process to fully succeed international initiatives providing for long-term financing and management of civil projects within Kazak military microbiological facilities are required.

1. Kazakstan's Network of BW Facilities: the Setting for Conversion

The key biological weapons facility in Kazakstan is the Kazakh Science Industrial Complex *Biomedpreparat* (*Kazakhstanskii nauchno-promyshlennyi kompleks "Biomedpreparat"* or *KNPK Biomedpreparat*) located around 10 kilometres from Stepnogorsk (Akmola *oblast'*). It was built some ten years after the signing of the Biological Weapons Convention and lay at the very heart of the secret Soviet offensive BW programme and has been described in recent US reports as the "main facility for the development and production of biological weapons in the Soviet Union" and the "central biological warfare centre of the Soviet Union"[2]. A recent Russian report described it as "one of the world's most powerful centres for the manufacture of bacteriological weapons" [3]. *Biomedpreparat* is reported to have carried out military research on toxins, viruses and bacteria [4].[1] It also embraces dedicated production facilities for weaponised forms of anthrax and other bacterial and viral agents.[2] A significant proportion of these facilities is located in reinforced concrete bunkers.[3] The eventual fate of this major military complex will be of critical importance in determining the success of international initiatives with regard to the dismantling of the Soviet Union's biological weapons programme.

Kazakstan also acted as a key staging post for access to the USSR Ministry of Defence's biological weapons test range on Vozrozhdenie Island in the Aral Sea. The first reported use of Vozrozhdenie Island (*Ostrov Vozrozhdeniya* or "Island of Rebirth") as a biological weapons proving ground was in the mid-1930s when it was

---

1. Very little is known about the precise nature of the work carried out although the facility has worked on the *Staphylococcus* enterotoxin A (SEA). Gertz [2] reports that *Biomedpreparat* has developed bioengineered weaponised strains.

2. One recent Russian report states that the anthrax produced by *Biomedpreparat* was resistant to a wide spectrum of antibiotics[3]. Participants at the ISTC seminar "Conversion in the Framework of International Collaboration" in Almaty (22-26 October 1996) confirmed that anthrax had been produced at Stepnogorsk.

3. A recent Russian report may have exaggerated when it referred to the fact that the "reinforced concrete hangars with two-metre thick walls are capable of withstanding a two kiloton nuclear strike." The same story claimed that underground passages link *Biomedpreparat* to Stepnogorsk and extend for "dozens of kilometres" in the vicinity of the facility [3].

utilised by the Red Army's Scientific-Medical Institute.[4] No further reference is made to the use of Vozrozhdenie for BW experiments until 1954 when it is reported that a secret base (known officially as Aral'sk -7) with a population of more than 1,000 was built on the island.[5] It utilised a biological weapons test range, which was initially 200 km^2 in size but had expanded to 2,000 km^2 by the 1990s. The proving ground was located on that part of the island, which belongs to the territory of Karakalpakstan, Uzbekistan. Military unit No. 25484 which was based on the island was part of, and subordinate to, a larger unit based in Aral'sk, Kazakstan, which itself received its orders from the USSR Ministry of Defence Institute of Microbiology's Centre of Virology in Sergiev Posad. Tests on the island have been linked to various environmental disasters in the region. It has been reported that two epidemics broke out on Vozrozhdenie during the late 1980s and early 1990s, the latter of which resulted in the deaths of approximately 434,000 animals. During the emergency the entire human population is reported to have been evacuated to Aral'sk-5 as a precautionary measure [7]. The island is still being linked to human illnesses in and around Muinok (*Muinak* in Russian) in Karakalpakstan, Uzbekistan [8].

A number of other microbiological facilities within Kazakstan were also integrated within the Soviet biological weapons programme. The Scientific-Research Agricultural Institute (*Nauchno-issledovatel'skii sel'skokhozyaistvennyi institut* or *NISKhI*) in Gvardeiskii, Otar, Dzhambul *oblast'* for example, was engaged in military research on a range of anti-plant and anti-animal agents [9]. Part of the institute's BW programme centred on animal viruses and among the agents studied have been Rinderpest, equine influenza, sheep pox and rabbit haemorhhagic fever.[6] Since the 1970s the institute has also been engaged in military research on the resistance of crops to various anti-plant biological agents. Another facility, which was previously focused on development of biological weapons for the Soviet military, is the Institute of Pharmaceutical Biotechnology (*Institut farmatsevticheskoi biotekhnologii*) in Stepnogorsk. The institute is a spin-off from *Biomedpreparat* and was originally the

---

4. This institute was later transferred to Kirov and is now known as the Russian Ministry of Defence's Scientific-Research Institute of Microbiology. Hirsch [5] reports that a 10,000 km^2 tract of land was allocated to Institute No. V/2-1094 (also known as the Velikanov Institute) which was a branch of the Red Army's Scientific-Medical Institute (also known as the Biotechnical Institute) in Moscow. According to him the resident population of Kulaks were driven off, and the first experimental expedition of 100 specialists and their support staff under Professor Ivan Mikhailovich Velikanov reached the island in the summer of 1936. Following the arrest of Velikanov and other specialists a second series of experiments, on a much greater scale, were initiated in the summer of 1937. The researchers were provided with special ships and two aircraft for their own exclusive use. The experiments were focussed on the dissemination of tularaemia and related microorganisms. Following an alleged breach of security the institute is reported to have been evacuated from the island in the autumn of 1937.

5. The decision to again use the island as a proving ground was made in 1952 but construction of facilities did not begin until 1954. It is reported to have incorporated a small landing strip, housing complex and laboratory [6].

6. The institute currently produces vaccines against sheep pox, Rinderpest, equine influenza, and rabbit haemorhhagic fever. Other product lines developed at the facility include a kit for diagnosis of sheep pox and African green monkey kidney tissue cultures and calf kidney tissue cultures. The Director, S. Mamadaliev, is reported to be one of the FSU's leading experts on animal viruses [10].

research arm of this BW facility.[7] Another Stepnogorsk plant with BW capability is the Joint Stock Company *Progress* (*Aktsionernoe obshchestvo Progress*). This facility, which is one of the largest microbiological complexes in the Former Soviet Union incorporating at least two production plants and 3,000 personnel, is located on territory contiguous with the BW facilities belonging to *Biomedpreparat.*[8] It was formerly subordinate to *Biopreparat.*[9]

The Kazak Scientific-Research Anti-Plague Institute (Almaty) also played an unknown role in the Soviet biological weapons programme.[10] The institute incorporates a network of seven anti-plague stations (including those in the Shymkent, Atyrau, Kzyl Orda, Mangistau, Taldy-Kurgan and Ural'sk *oblasts*), 18 anti-plague departments, laboratories for the production of vaccines, sera and diagnostics, and facilities which provide specialist training for physicians, biologists and laboratory assistants.[11] It has sufficient capacity to turn out twenty-one million doses per annum of a live dry plague vaccine (based on the EV strain) [14]. Other microbiological facilities in Kazakstan which may have been linked to the defence complex include the M.A. Aitkhozhin Institute of Molecular Biology & Biochemistry (*Institut molekulyarnoi biologii i biokhimii imeni Aitkhozhina M.A.*) in Almaty, the Institute of Physiology, Genetics and Bioengineering of Plants (*Institut fiziologii, genetiki i bioinzhenerii rastenii*) in Almaty and the Almaty Biocombine (*Almatinskii biokombinat*).[12] The latter has large-scale capacity for production of anthrax, brucellosis and foot and mouth disease vaccines.[13]

---

7. Speakers at the recent ISTC meeting in Almaty referred to the fact that the institute used to be a component part of *Biomedpreparat* which was itself engaged in the research, development and production of biological weapons [11].

8. The Joint Stock Company incorporates at least two production plants, the *Biopreparat* Factory and the *Aminokisloty* Factory [12].

9. According to a recent US report *Progress* formerly had a defence affiliation and possesses the capability to produce BW agents [4].

10. In 1997 it forwarded a project application to the International Science & Technology Centre (ISTC) in which it was stated that 126 "persons-months of effort of weapon scientists and engineers" would be employed. The Project is entitled "Mechanism of Formation of Pathogenic Bacteria Virulence Population Level. Application of Pathogenic Bacteria Cytotoxic Impact on Cellular Single Layer as a Test-System to Determine the Pathogenicity (Virulence Level) of Bacteria Strain in Vitro".

11. A list of six anti-plague stations which were to be transferred to the control of the Kazak Scientific-Research Anti-Plague Institute were given in a Ministry of Health publication in 1992. However, Alim Masgutovich Aikimbaev (Deputy Director) reported in April 1995 that seven stations were subordinate to his institute. The seventh station might be that attached to the *Yuzhpolimetal* Production Association (location unknown) [13].

12. The Almaty Biocombine was described as a "conversion enterprise" by Gennadii Nikolaevich Lepeshkin (head of the Kazak National Centre for Biotechnology) at the ISTC seminar in Almaty [15]. Lepeshkin has also reportedly stated that all facilities now incorporated within the National Centre for Biotechnology were part of the defence complex. This would therefore appear to suggest that the Aitkhozhin Institute and the Institute of Physiology, Genetics and Bioengineering of Plants had some connections with military programmes [16].

13. The facility appears to have formerly been subordinate to the USSR Ministry of Agriculture via the Main Administration of the Biological Industry (*Glavbioprom*). It was the Soviet Union's second largest vaccine production facility (only the Shchelkovo Biocombine in Moscow had larger capacity). It produces at least four anthrax vaccines including the *STI* (*Sanitarno-tekhnicheskii institut*) anthrax vaccine and the *VNIIVViM* vaccine based on the naturally attenuated anthrax strain No. 55 [17].

2. The Development of BW Facilities at Stepnogorsk

A number of factors influenced the Soviet authorities in their decision to locate a major new BW facility in Northeast Kazakstan including its extreme remoteness but also the availability of local utilities developed by the Soviet defence complex. The town of Stepnogorsk itself is located two hundred kilometres to the North of Akmola, close to the border of the adjacent Kokchetav *oblast'*. The town originated as a result of the opening of a new mining and chemical combine by the USSR Ministry of Medium Machine Building (*Minobshchemash*) in 1956.[14] *Minobshchemash* was part of the Soviet defence complex and was responsible for production of strategic and intermediate range ballistic missiles [18]. Stepnogorsk was formerly a "closed" town and remains a major R & D and production centre for the military-industrial complex in the former Soviet Union. Before the collapse of the USSR it was referred to under several names including "Makinsk-2", "Tselinograd-25" and "Aksu". In June 1964 an unpublished decree of the Presidium of the Supreme Soviet of the Kazakh SSR conferred the status of a town upon Aksu and it was given the name Stepnogorsk.[15] The domination of Stepnogorsk by *Minobshchemash* is of particular interest since a recent report has suggested that the management structure of the State Concern *Biopreparat* (responsible for the illicit Soviet offensive biological weapons programme) was an offshoot of this Ministry.[16] Whether this is the case or not, the origins of the microbiological complex in Stepnogorsk do appear to lie in the close personal relations which existed between Vasilli Dmitrievich Belyaev, the Head of the Main Administration of the Microbiological Industry (*Glavmikrobioprom*), and Efim Slavskii, Minister of *Minobshchemash*. It was Slavskii who made available *Minobshchemash*'s excess construction capacity to *Glavmikrobioprom* for the erection of its first factory in Aksu [20].

Construction of the buildings comprising the Kazakh Science Industrial Complex *Biomedpreparat* began at Stepnogorsk in 1982. The facility was originally a branch of the Moscow-based Scientific-Research Institute *Biokhimmashproekt*, (*Kazakhskii filial nauchno-issledovatel'skogo instituta "Biokhimmashproekt"*) and was ultimately subordinate to the State Concern *Biopreparat*.[17] Some time after 1982 it was renamed

---

14. The Minister of *Minobshchemash*, Efim Pavlovich Slavskii, personally selected the site for the construction of Stepnogorsk.

15. Among the other production plants which have since been built at Stepnogorsk are a Preserved Milk Products Combine, a Bearings Factory and a Ceramics Factory [19].

16. According to Leitenberg *Biopreparat*'s management staff were taken from the Ministry of Medium Machine Building. Leitenberg, M. (1994) [1].

17. The Scientific-Research Design Institute of Applied Biochemistry (*Biokhimmashproekt*) is based at Ulitsa Klary Tsetkin, 4/6 in Moscow. Its Director in 1993 was V.G. Popov. The institute is currently engaged in the manufacture of various types of equipment including that used for microfiltration, sublimation and spraying. It is not known if close links still exist between this facility and its former branch in Kazakstan. The institute is also the head organisation of the Science Production Association *Biomash* (*NPO Biomash*). *NPO Biomash* has supplied fermenter installations and other technology to factories located at Pokrov, Omutninsk and Stepnogorsk. In January 1993 it was reported that a 1,000 m^2 production facility called "Building D" which

the Stepnogorsk Scientific Experimental-Industrial Base (*Stepnogorskaya nauchnaya opytno-promyshlennaya baza* or *SNOPB*) [22]. *SNOPB* was reorganised as the Kazakh Science Industrial Complex *Biomedpreparat* in 1993 [23]. The *Biomedpreparat* complex presently embraces 25 buildings and is reported to occupy a site 43 hectares in size. The following units have been identified at the site: Building 211 (Food media production); Building 221 (Main Biotechnology Facility); Building 231 (Vacuum drying of biological preparations); Building 241 (Production of immobilised enzymes and ginseng); and Building 600 (Experimental Building possessing lines for tableting and ampoule and vial filling). Other buildings are concerned with power supply, steam generation etc. In addition *Biomedpreparat* maintains a vivarium [24].

One of the pivotal figures in the development of the Stepnogorsk BW facilities was Kanatzhan Baizakovich Alibekov, a relatively youthful Kazak, who was appointed Director of the Stepnogorsk Scientific Experimental-Industrial Base in 1983.[18] He was transferred to *SNOPB* from a *Biopreparat* facility which he headed in Omutninsk (Kirov *oblast'*) and where he had been recognised as "an energetic manager with great prospects".[19] Some time around 1987-1988 he left Stepnogorsk and was appointed as Deputy Director of the *Biopreparat* network with responsibility for its scientific programmes.[20] It was Alibekov who in the early 1990s took the initiative in developing the first commercial contacts between *Biopreparat* and Western biotechnology companies.[21] However, Alibekov also played a vitally important role in bringing about the downfall of the Soviet offensive BW programme since he subsequently had the distinction of becoming the highest ranking defector from *Biopreparat* to the CIA and is presumably the "very senior official" referred to in Western intelligence reports.[22]

---

originally belonged to *Biomash* was seized by a group of men wearing paratrooper uniforms and armed with police truncheons. They were acting on behalf of the Joint Stock Company *Apromed* which claimed that the facility had been legally privatised and belonged to them. Despite the fact that the State Concern *Biopreparat* had declared *Apromed*'s original acquisition of this facility to be illegal, the Joint-Stock Company planned to demolish part of the fence surrounding *NPO Biomash* and to create a gate allowing access to "Building D". Such incidents highlight the chaotic environment in which the FSU's former BW facilities exist today [21].

18. Eduard Ivanovich Perov was the Director of the facility during the construction phase (he was subsequently appointed head of the *Progress* facility and remains in that post today). Perov was replaced as Director for a short period by Yurii Petrovich Davydkin before Alibekov's arrival at the end of 1983.

19. While at Omutninsk Alibekov prepared his Candidate of Science degree under the supervision of Major-General Anatolii Andreevich Vorob'ev. Although considered weak his dissertation was accepted because there was considerable sympathy for him having completed it in such a remote area of the Soviet Union [25].

20. Alibekov arrived in Moscow having just completed his doctoral dissertation. According to one expert it was definitely not of a high enough standard and was really only a description of a simple experimental-industrial procedure. However, the head of *Biopreparat*, Yurii Tikhonovich Kalinin, the head of the Ministry of the Medical & Microbiological Industry, Valerii Alekseevich Bykov and other senior officials successfully lobbied for the doctorate to be awarded [25].

21. In 1990 for example Alibekov undertook negotiations with Derek Lennon, Chairman of Oxford Virology, concerning the formation of a joint venture which would comprise of the UK company, Moscow Narodny Bank, the State Concern *Biopreparat* and the Science Production Association *Vektor*. The aim of the joint venture would be to "market services (and) promote into the world market the products and technologies of *Biopreparat* subordinate facilities and research and development centres." Unfortunately Oxford Virology ceased trading before negotiations concerning the creation of the new venture were complete [26].

22. Alibekov is reported to have known significantly more about the *Biopreparat* BW network than the earlier defector Vladimir Pasechnik who reached the UK in 1989. According to one senior Russian source associated

3. The Reorganisation of Military-Microbiological Facilities in Kazakstan

The BW facilities at Stepnogorsk represent by far and away the most advanced microbiological production capability within Kazakstan. In 1993 the Kazak government took the decision to make *Biomedpreparat* and the associated facilities at Stepnogorsk the hub of a new civil biopharmaceutical industry and under the terms of decree No. 1140 issued by the Kazakh Cabinet of Ministers on the 16 November the National Centre for Biotechnology of the Republic of Kazakhstan (*Natsional'nyi tsentr po biotekhnologii Respubliki Kazakhstan* or *NTsB* for short) was created [28].[23] *NTsB* is based in Stepnogorsk and incorporates a number of former BW-related facilities including: the Kazakh Science Industrial Complex "*Biomedpreparat*" (Stepnogorsk); the Biocombine (Almaty); the Scientific-Research Agricultural Institute (Gvardeiskii, Dzhambul *oblast'*); the Institute of Pharmaceutical Biotechnology (Stepnogorsk); the M.A. Aitkhozhin Institute of Molecular Biology and Biochemistry (Almaty); the Institute of Physiology, Genetics and Bioengineering of Plants (Almaty); the Joint Stock Company "*Progress*" (Stepnogorsk); and the Medico-Biological Institute (Stepnogorsk) [30]. A branch headquarters of *NTsB* has been created in Almaty to facilitate communications with international biopharmaceutical companies.[24]

NTsB has been made the head organisation (*golovnaya organizatsiya*) of a new national civil biotechnology programme within Kazakstan. The Republican Targeted Scientific & Technical Programme (*RTsNTP*) is entitled "The utilisation of biotechnological and genetic engineering methods in medicine, agriculture and industry" ("*Ispol'zovanie metodov biotekhnologii i gennoi inzhenerii v meditsine, sel'skom khozyaistve i promyshlennosti*") and it incorporates more than two hundred individual scientific projects being conducted at twenty-five R & D facilities [17]. One of the main aims of the programme is the conversion of Kazakstan's military microbiological facilities but it is also charged with: the construction of modern GMP standard biopharmaceutical production facilities; the determination of R & D and production priorities within the bioindustry; the provision of support for fundamental research; the initiation of training schemes to provide a new generation of specialists for the industry; and the development of international contacts and collaboration [10]. A number of research areas have been selected for priority funding and these include

---

with the offensive BW programme Pasechnik's defection occurred at a time when it was considered "a betrayal of the motherland". Alibekov, on the other hand, "did it later without any fuss. The country was under a new power and the country was different" [25]. Further details of the defections from *Biopreparat* are described in two recent publications describing the activities of British intelligence [27].

23. Following the break-up of the Soviet Union and prior to the creation of *NTsB* a substantial section of the Kazak biopharmaceutical industry had been transferred in December 1991 to the control of the *Kazfarmbioprom* Concern (based in the city of Shymkent) under the Kazak Council of Ministers. *Kazfarmbioprom* incorporated eleven facilities including the *Progress* plant, the Almaty Biocombine and the Factory of Feed Antibiotics (Turkestan). However, the axis of power within the industry has now shifted to Stepnogorsk and the Shymkent organisation has had its role substantially downgraded [29].

24. The branch (filial) is located at the Ministry of Sciences-National Academy of Sciences' Institute of Experimental Biology on Prospekt Abaya in Almaty [31].

work on development of genetic engineering methods and the creation of gene banks and collections of plant and animal cells, and microbial and viral strains. The Baikonur Cosmodrome (Tyuratam), the site of all the Soviet and Russian manned flights, will also be utilised for the development of space biotechnology.

Side by side with the creation of *NTsB* the Kazakstan Cabinet of Ministers issued a decree on the 18 November 1993 "Concerning measures for the development of the pharmaceutical industry of the Republic of Kazakstan". The decree envisaged the creation of a new department within the Kazak Ministry of Health (*Minzdrav*) which would assist in the conversion of facilities in Stepnogorsk and the reconstruction and modernisation of biopharmaceutical facilities across Kazakstan [33]. In August 1995 a new department subordinate to *Minzdrav*, the Main Administration for the Development of the Pharmaceutical & Medical Industry (*Glavnoe upravlenie po razvitiyu farmatsevticheskoi i meditsinskoi promyshlennosti*) was duly established. Its new head, Marat Daleev was given the task of bringing on stream four new plants and developing Kazakstan's existing biopharmaceutical facilities [34].

The Kazak government has also implemented major changes in the management and organisation of the country's network of anti-plague stations. Following the break-up of the Soviet Union all plague prevention facilities in Kazakstan were transferred from the USSR Ministry of Health to the control of the Kazakstan Ministry of Health [35]. In addition the Kazak Ministry of Science-Academy of Sciences may now have some managerial role [36]. Under the auspices of the Ministry of Health the Kazak Scientific-Research Anti-Plague Institute in Almaty has become the head organisation of a new Science Production Association of Anti-Plague Establishments of the Republic of Kazakstan (*Nauchno-proizvodstvennoe ob"edineniya protivochumnykh uchrezhdenii Respubliki Kazakstan*). The new Association has responsibility for the management of anti-plague stations, departments and vaccine production laboratories across Kazakstan [37].

The other key initiative, which is having a major impact on the management and scientific programmes at Kazakstan's BW facilities, is the government's mass privatisation programme. It appears to be the case for example that the Joint Stock Company *Progress* (*AO Progress*) is to be broken up into three core businesses.[25] The first of these, *AO Progress* (with Valerii Anatolevich Shimanaev remaining in post as Director) will retain the bulk of the facilities and personnel (2,621 employees). The two much smaller spin-off companies will be *AO Premiksy* (headed by Vladimir Fedorovich Astashov with 144 personnel) and *AO Gerbitsidy* (headed by M.F. Subkhankulov with 63 personnel) [39]. In April 1996 a decree was issued outlining the sale of state blocks of shares in these companies by the State Management Committee and the State Property Committee subject to state approval [40]. As is the case with *Progress*, *Biomedpreparat* already has the status of a joint stock company with a recent report revealing that the state has retained 50 per cent of ownership in this key BW facility, but that 40 per cent had been transferred to Astyk, a state-owned agricultural

[25]. The Joint Stock Company "*Progress*" was created on the basis of the Production Association "*Progress*" in 1993 [38].

holding company, and 10 per cent to the employees themselves [4]. However, this facility too is scheduled for privatisation under the terms of the April 1996 decree.

The major success story thus far with regard to the dismantlement of Kazakstan's offensive BW capability is the closure and dismantling of the proving ground on Vozrozhdenie Island and the associated unit located at Aral'sk. In 1991 the base's personnel were evacuated and the laboratories mothballed. Subsequently in November 1991 at a meeting of the Scientific Council of the USSR Ministry of Defence Institute of Microbiology's Centre of Virology in Sergiev Posad the decision was taken to terminate experimental work on the island. Work on dismantling the base is then reported to have begun in May 1992 [41]. US officials later confirmed that the test site had been closed and was being decontaminated.[26] However, cynics might argue that the military withdrawal from the island was prompted not so much by obligations under international treaty but rather by impending environmental changes. Projections suggest that Vozrozhdenie is likely to disappear as an island by the year 2000 and will be joined to the mainland by a substantial land bridge.[27]

4. Civil R & D and Production Programmes within Kazakstan's BW Facilities

Following the break-up of the Soviet Union the Kazak government has established civil R & D and production programmes at its military microbiological facilities. In 1991 *Biomedpreparat* began industrial production of an immobilised protease called "*Profezim*" [22]. This preparation was originally developed by researchers at the Russian Academy of Sciences' Institute of Cytology & Genetics (Novosibirsk) and the State Concern *Biopreparat*'s Scientific-Research Design Institute of Biologically Active Substances (Berdsk) [44]. *Profezim* accelerates the wound-healing process and has been found to be especially effective in the treatment of burns, festering wounds, trophic and radiation ulcers and chronic osteomyelitis. One of the key customers for the preparation in the recent past has been the USSR Ministry of Defence who used it in the treatment of battlefield casualties [45]. Presently demand for *Profezim* and analogous preparations significantly exceeds supply and plans have been drawn up for the establishment of an independent production unit at *Biomedpreparat*. However, until recently a key limiting factor for the expansion of production has been the lack of specialised sterilisation equipment and consequently immobilised enzymes have had to be sent to Moscow for treatment. In order to address the problem of sterilising these preparations and other pharmaceuticals and medical products (such as disposable syringes) an important new project was initiated at the complex. The construction of a

---

26. General Anatolii Kuntsevich is also quoted in the same report as stating that the material base for weapon testing had been destroyed and that the island would be safe for civilian use in two to three years [42].

27. Presumably the original raison d'être for selection of Vozrozhdenie include both its geographical remoteness from centres of population, and the fact that as an island it was better suited, in theory at least, to the containment of disease outbreaks. However, between 1966 and 1993 the Aral Sea sunk by more than 16 metres, its total volume was reduced by 75 per cent and its area by about half. Present projections reveal that part will continue to shrink and that by the year 2000 Vozrozhdenie is likely to disappear as an island and will be joined to the mainland by a substantial land bridge [8, 43].

radioisotopic sterilisation building, which uses cobalt-60 for irradiation with γ-rays, was completed in 1994. It is intended that the new facility will be operated in conjunction with the *Medpolimer* enterprise in Pavlodar [22].

Biomedpreparat has since developed its civil manufacturing capability and is presently engaged in the manufacture of a diverse range of products. It is now Kazakstan's major producer of nutrient media. Production of 17 different types of media have been assimilated at the facility which has a total capacity of 6,000 tonnes per annum [24]. These products are supplied to microbiological research institutes, hospitals and epidemiological stations across the country [22]. Media is also exported to Novosibirsk (Russian Federation) and the Baltic States [24]. A number of vaccine preparations are also produced by *Biomedpreparat* including a BCG (anti-tuberculosis) vaccine and a brucellosis vaccine.[28] Production of medical and veterinary lactobacterin and bifidumbacterin has also begun.[29]

However, a major constraint upon the development even of the somewhat limited conversion programmes thus far implemented has been the lack of adequate financing from the Kazak government. Until the advent of recent international projects *Biomedpreparat* had been in the grip of a semi-permanent financial crisis. Funding from Moscow was suspended following the break-up of the Soviet Union and it was not until the end of 1992 that the Kazak government provided an emergency payment of 50 million roubles to the facility. This was followed by an additional 90 million roubles funding for the first quarter of 1993 [47]. However, the situation again deteriorated and in July 1995 it was reported that even salaries to senior scientific personnel had been left unpaid for several months [48].

A recent project initiated at *Biomedpreparat* has been the reequipment of existing facilities for the manufacture of disposable syringes. It was intended that the first production line would start up in 1994 with an annual capacity of 25 million syringes. Subsequently it was intended that production would be scaled up to 150 million syringes per annum [22]. However, delays in financing appear to have put this work behind schedule and production does not yet appear to have been initiated. Again in September 1994 it was reported that Kazak scientists had prepared plans to open a new facility at *Biomedpreparat* for the production of drugs from plants. Sufficient capacity was reported to exist but funds would be required to reequip and convert part of the enterprise. Apparently, Lepeshkin, Marat Sovetovich Koishibaev, the Head of Administration of Stepnogorsk (*Stepnogorskaya goradministratsiya*), and Shimanaev, the President of the Joint Stock Company *"Progress"* were all receptive to the idea but once again the project floundered because of a lack of adequate financing. More recently in 1996 representatives of *Biomedpreparat* contacted a UK distributor with a

---

[28]. A recent US report states that since 1991 the facility has been producing a vaccine against plague. The only reference to a product such as this in the Kazak literature is a medicinal-prophylactic sera against canine distemper (in Russian "canine plague") [4, 20].

[29]. Other products manufactured by the facility in the past have included thymidine (isolated from DNA in fish roe or bacteria) and recombinant alpha-2 interferon (marketed under the name "*Reaferon*"). The facility is also reported to have produced the preparation *Apivit* used for the therapeutic treatment of bees. In 1990 300 litres of the preparation were delivered to Georgia and the Baltic States also imported large quantities. However, there was minimal demand for the product in Kazakstan [22, 46].

request that they source second-hand tablet presses with a total value of $100,000. However, when the equipment was located and offered to the Kazaks they were unable to proceed with the purchase because the promised line of funding did not materialise [49].

Perhaps the key conversion project to have been stymied in recent years has been that concerning the reconstruction of Building No. 221. This facility which has been described as the heart of the military complex at Stepnogorsk was constructed in 1982 and houses a production area subject to an unknown level of containment. Lepeshkin and his colleagues wished to utilise the facility for the production of recombinant human insulin based on a process originally developed in Moscow and subsequently modified in Stepnogorsk. In 1991 *Biomedpreparat* succeeded in scaling-up the process technology to $20m^3$.[30] Government support for the project was first pledged during a visit by President Nazerbaev. This was followed by publication of a special decree of the Kazak Cabinet of Ministers on "The Manufacturing of Medical Preparations in Stepnogorsk" which earmarked US$25 million investment for *Biomedpreparat*, the bulk of the money intended for the insulin project [24]. Lepeshkin began moves to procure a complete GMP production line, including a fermentation plant and down-stream processing facilities. However, despite his personal commitment to the scheme and the establishment of a line of credit in a European country *Biomedpreparat* have been unable to secure the initial investment from the Kazak government to initiate the project. This and the other examples detailed above vividly illustrate the present limitations of conversion programmes implemented and funded from within Kazakstan and point to the need for long-term international financing and management of civil projects within Kazak military microbiological facilities.

5. International Concerns: Security & Controlling the "Brain Drain"

Two of the principle concerns of Western governments with regard to Kazakstan's BW network are the security of individual facilities and the stocks of microbiological pathogens and toxins which they might maintain, and the potential problem of foreign governments using financial and other inducements to recruit disgruntled BW scientists and engineers. The first of these concerns has been thrown into sharp focus by two recent reported incidents, which appear to suggest a lack of adequate security precautions to prevent unauthorised access to Kazakstan's collections of plague strains and other highly pathogenic microorganisms. In 1995 it was reported that cultures had been stolen from a Kazak anti-plague institute and that criminal gangs planned to disseminate plague, cholera and anthrax in the Russian city of Khabarovsk. The city's police force allegedly undertook "emergency safety measures" in response [50]. It would also appear to be the case that the Kazak anti-plague network has been attempting to commercially exploit its collections of plague strains. According to an

---

30. The process was developed by Sergei Mikhailovich Navashin's Institute of Antibiotics in Moscow. According to Lepeshkin the Russian process possesses economic advantages over that developed by Eli Lilly and the number of stages have been reduced.

English-language paper in Delhi a company called Viva based in Almaty has produced a catalogue with entries on 29 different strains including "high virulent mutant", "natural high virulent" and "natural avirulent" and includes details of their resistance or sensitivity to different antibiotics. The existence of such a company has again raised international concern that pathogenic microorganisms could be procured by members of terrorist groups or countries, which are actively pursuing an offensive biological weapons programme.[31]

It is difficult to estimate the precise number of specialised biological weapons scientists and engineers who were employed on the Soviet BW programme at the *Biomedpreparat* site. In 1984 it is reported that 350 personnel were employed at the facility but many of these were newly graduated specialists who possessed very little practical work experience. The number of staff working at *Biomedpreparat* doubled over the next seven years or so and appears to have peaked at around 800 in 1991 with this total including one Doctor of Science, fifteen Candidates of Science and 50 research associates [46]. By August 1994 the number of personnel had been reduced by 50 per cent to give a total workforce of 450 people[48]. However, despite this reported cutback following the suspension of Soviet funding, it does appear that at least up until this time *Biomedpreparat* had been successful in retaining its most highly qualified personnel and the workforce at this time still included fifteen Candidates of Medical, Biological or Technical Sciences.[32] However, more recent reports suggest that there has been a considerable brain drain at *Biomedpreparat*, which has lost the bulk of its qualified scientific staff and engineers and that of the remaining 300 staff only around sixty are key personnel [52]. Although the government's biotechnology programme envisages a better rate of remuneration for scientists employed within its microbiological facilities there is no evidence that significant increases in salaries have thus far been implemented. In an urgent attempt to generate replacements for the specialists who have departed, the National Centre for Biotechnology has established a new specialised training college in Stepnogorsk, the Medico-Biological Institute (*Medikobiologicheskii institut*). It was opened in 1994 and has an annual intake of seventy students who are recruited from both Kazakhstan and other CIS states. The first students will graduate from the institute in 1998.[33]

It is not possible to be certain on what has happened to the scientists and engineers who have left Stepnogorsk but what little evidence, which is available,

---

31. The scientist with principle responsibility for the strains in the catalogue is Professor I.L. Martinevskii although O.V. Struchkova, E.N. Struchova, A.I. Kotova and V.M. Stepanov are also credited with development of strains in the catalogue [51].

32. The Candidates of Medical, Biological or Technical Sciences at this time were: Rumyantsev, V.A.; Mikhailichenko, A.P.; Narozhnov, V.V.; Petrov, P.P.; Brezitskii, V.B.; Grechanyi, L.A.; Toropova, N.E.; Staritsyn, N.A.; Tolstykh, T.M.; Zakhryalov, S.Ya.; Khaitonov, V.G.; Polozhentsev, S.I.; Skitev, A.V.; Bugreev, V.N.; and Bai, L.A. [22].

33. There are three faculties devoted to microbiology, biochemistry and business management (in biotechnology). Students enter at age 17 and are enrolled for five years. During the last two years of their course they are given on-site training at facilities belonging to both the Joint Stock Company *Progress* and the Science Industrial Complex *Biomedpreparat*. The ultimate objective is that graduates will find employment in both *Progress* and *Biomedpreparat*. The college also possesses its own IT training department which possesses twenty new computers donated by the Kazak government [53].

suggests that the majority have probably found alternative employment within the Russian biopharmaceutical industry.[34] Nevertheless, the possible transfer of BW personnel from Kazakstan to countries which are pursuing their own programmes for the development of biological weapons remains a matter of major concern to Western governments and has prompted several international initiatives aimed at supporting microbiological facilities in Stepnogorsk and integrating them within the global business and scientific community.[35]

6. Integrating Kazakstan's BW Facilities within the International Scientific Community

There have been a number of initiatives originating both within Kazakstan and from external agencies aimed at the integration of the country's BW facilities within the international scientific community. One of the earliest moves in this direction was the organisation by Genadii Lepeshkin (head of *NTsB*) in August 1995 of an international scientific conference in Stepnogorsk, which incorporated visits to BW and civil biopharmaceutical facilities. However, although representatives from a wide range of Russian research facilities linked to the defence complex gave presentations there appears to have been no substantive international scientific input or representation at this event.[36]

A more serious attempt at constructive engagement with Kazak military microbiological facilities has been via the International Science & Technology Centre (ISTC) which is an intergovernmental organisation based in Moscow comprising the European Union, USA, the Russian Federation, Japan and eight other countries. It seeks to provide funding for former weapons scientists and engineers engaged in Soviet programmes aimed at creating weapons of mass destruction, to work on civil projects. Kazakstan became a member of the ISTC at the sixth Board meeting held on 29-30 June 1995 [57]. Since then several projects have been approved or are pending with Kazak microbiological facilities [58]. A strict requirement of ISTC funding is that an active Western partner is required for any project it funds in order to ensure that the work is directed at civil applications, thereby also helping to ensure that linkages are made between the Kazak weapons scientists and the international scientific community.

The most important international initiatives with regard to the dismantling and conversion of Kazakstan's BW capability have originated from the USA. Recognising

---

34. At least two personnel from Stepnogorsk for example have recently found employment at *Rosagrobioprom*'s Kursk Biofactory [54].

35. There is no evidence available with regard to foreign recruitment of BW scientists but a number of countries have sent delegations to assess Kazakstan's biopharmaceutical capability including Israel and Kuwait [55].

36. Among the centres represented at the conference were three *Biopreparat* facilities, the State Scientific Centre of Applied Microbiology (Obolensk), the State Scientific Centre of Virology & Biotechnology *Vektor* (Kol'tsovo) and the Institute of Biologically Active Substances (Berdsk). The Russian Ministry of Defence's Institute of Microbiology (Kirov) was also represented. The proceedings of the conference are published in five volumes [56].

the central role played by *Biomedpreparat* within the Soviet BW programme US agencies have devoted considerable resources to assisting in the identification of a new role for the facility within Kazakstan's civil biopharmaceutical industry. *Biomedpreparat* has been included within the Nunn-Lugar Co-operative Threat Reduction project "Weapons of Mass Destruction Infrastructure Elimination" (WMDIE). It is envisaged that the Biological Weapons Production Facility Dismantlement Project will be accomplished in two distinct phases. The first entailed a contract with *Biomedpreparat* to develop a Management Action Plan covering the work to be accomplished, the identification and itemisation of equipment and systems that must be removed to eliminate BW production capability, the sampling of areas key to production of biological weapons, the establishment of a US-equipped safety laboratory and the development of a dismantlement plan. The first phase was initiated on the 5 December and was scheduled for completion in August 1997. It is envisaged that the second phase will focus on the physical dismantlement of the specialised equipment and critical systems to ensure the demilitarisation of the facility.[37] However, this second crucial part of the project may be facing some opposition and critics have already voiced concerns that once the US have fulfilled their basic mission, the destruction of the military infrastructure, then they will cease to invest in the construction of new plant for the production of biopharmaceuticals [3].

The criticism of the WMDIE project follows the reported collapse of the first commercial conversion project at Stepnogorsk [60]. Under the umbrella of the Co-operative Threat Reduction Programme *Biomedpreparat* was selected as a participant in an industrial partnerships scheme. The US Defence Special Weapons Agency (DSWA), which was responsible for procurement of appropriate Western companies for participation selected Allen & Associates International Ltd. (AAI) based in Washington as *Biomedpreparat*'s partner. DSWA awarded a $5.82 million cost-sharing contract with the US contributing $2,685,992, and AAI an additional $3,139,620 [61]. The collaborative project was centred on a joint venture registered in March 1996, KAMED Resources. It aimed to utilise 1-2,000m^2 of a facility previously utilised for the production of biological weapons for the manufacture, packaging and distribution of vitamins, pharmaceuticals and antibiotics.[38] In May 1996 the project was given a major boost when the international company ICN Pharmaceuticals, Inc. (USA) linked up with AAI in the conversion of the Kazak plant.[39] However, despite this crucial injection of industrial expertise it has recently been reported that the first project aimed at the production of antibiotics collapsed when manufacturing equipment

---

[37]. The original WMDIE agreement was signed on 3 October 1995 by the US Department of Defence and the Kazak Ministry of Science - Academy of Sciences which provided an initial $6 million for the elimination of nuclear test tunnels at Semipalatinsk. An amended agreement was signed between these two organisations on the 7 June 1996 providing additional funding to a total of $22.5 million and permitting usage of WMDIE funds for closure of nuclear test holes at Semipalatinsk and the dismantlement of the infrastructure for BW production at Stepnogorsk [59].

[38]. Details of US investment in *Biomedpreparat* were announced by Harold Smith (Special Assistant to Defence Secretary William Perry) at a news conference in April 1995 in Almaty with Mr Perry and Sagdat Nurmagambetov (Kazak Defence Minister) [2].

[39]. In addition to financial support ICN were to contribute equipment, materials expertise and training [62].

supplied from the US turned out to be hopelessly obsolete. Whatever the current status of the project this episode does clearly illustrate the difficulties involved in development of commercially viable biopharmaceutical companies able to produce pharmaceuticals to acceptable international standards [3].

7. Issues Outstanding

In what is a very delicate political situation it is still not absolutely clear whether there has been a total and irreversible transfer of control of BW facilities in Kazakstan from Moscow-based organisations to civil bodies under the jurisdiction of the Kazak government. Although Kazakhstan in a strictly legal sense now has full control of these facilities does the Russian Federation still play a major role in key decision-making at Stepnogorsk? Following the collapse of the Soviet Union, *Biomedpreparat*'s links to Moscow and *Biopreparat* were formally broken. However, the situation on the ground is a little more complicated than may at first glance appear. Stepnogorsk is located in an area of Kazakstan, which is populated in the main by ethnic Russians who still feel a strong bond of loyalty to the Russian Federation [63]. The region's ethnic balance is repeated within the town itself with Russians comprising 79.2 per cent of the population and Kazaks only making up 2.7 per cent of the total inhabitants.[40] In addition *Biomedpreparat* and associated facilities are managed and staffed by personnel who have in the very recent past been closely linked to the Russian military-industrial complex. Gennadii Nikolaevich Lepeshkin for example, was one of a number of senior Soviet military scientists transferred from microbiological facilities subordinate to the USSR Ministry of Defence to senior positions within the *Biopreparat* organisation. Before taking up a management position at *Biomedpreparat*, and subsequently the headship of the National Biotechnology Centre[41], Lepeshkin was employed at the Ministry of Defence's Scientific-Research Institute of Microbiology in Kirov, Russia.[42] Other Russians holding key positions in Stepnogorsk include Yurii Aleksandrovich Shapovalov, (a Deputy Director of *NTsB*) [30], Aleksandr Nikolaevich Kosinov (former Deputy Director and Head of Department of *Biomedpreparat* and now Director of the Institute of Pharmaceutical Biotechnology) and Aleksandr Petrovich

---

40. Stepnogorsk's population comprises 74 nationalities including: 79.2 per cent Russians; 2.7 per cent Kazaks; 7.3 per cent Ukrainians; 3.7 per cent Germans; 2.5 per cent Tartars; 1.4 per cent Belarussians; and other nationalities 3.2 per cent (see *Akmola entsiklopediya* [52]).

41. Interestingly, Lepeshkin replaced the Kazak, Professor Murat Kurmashevich Gilmanov as the head of *NTsB*.

42. Lepeshkin was born in the Murmansk *oblast'* in 1946. He graduated from the Gor'kii Medical Institute in 1962. His first appointment was as a Junior Scientific Worker at the Institute of Microbiology in Kirov. He has worked in Stepnogorsk since 1984. Initially he was Head of a Department at the Stepnogorsk Scientific Experimental-Industrial Base (*Stepnogorskaya nauchnayaopytno-promyshlennaya baza* or *SNOPB*). In 1987 he was appointed head of this organisation and remained in this position when it was reorganised as the Kazakh Scientific-Industrial Complex *Biomedpreparat* in 1993. His most recent appointment has been as the new General Director of the National Centre for Biotechnology of Kazakhstan. He is a Doctor of Medical Sciences, a Professor and the author of 140 scientific papers and inventors certificates [23].

Mikhailichenko (who was promoted from Deputy Director to Director of *Biomedpreparat*).[43]

There is also some evidence to suggest that the Joint Stock Company *Biopreparat* continues to play an important role with regard to the continuing development of facilities in Stepnogorsk. In 1993 for example, *Biopreparat*'s State Institute for the Design of Enterprises of the Biological Industry (*Giprobioprom*) based in Moscow was reported to have been engaged in the reconstruction of *Biomedpreparat*'s Building 221 where it is planned to begin production of recombinant human insulin using imported Western equipment.[44] This institute, which also has a branch based at Kol'tsovo, is one of *Biopreparat*'s main design facilities [64]. *Biopreparat*'s role in the upgrading of facilities at Stepnogorsk is not at all surprising given that it developed the original production technology at the site and that the economies of Russia and Kazakstan remain closely integrated. However, this example does illustrate the point that Moscow-based organisations such as *Biopreparat* which were integral parts of the Soviet offensive BW programme continue to be able to operate beyond the boundaries of the Russian Federation and may have continued access to numerous microbiological R & D and production facilities in the newly independent states.

There have also been a number of initiatives aimed at the reintegration of Kazakstan's BW facilities within joint CIS biotechnology programmes coordinated from Moscow. At a meeting in August 1995 for example leading figures from the biopharmaceutical industries of the Russian Federation, Belarus, Ukraine and Kazakstan (represented by Gennadii Lepeshkin) met in Moscow to discuss collaboration in fundamental and applied R & D and production. Similarly in May 1997 at a special meeting of the Russian Ministry of Health's Collegium it was argued that a single centralised anti-plague system for the Former Soviet Union as a whole should be established which would include the large network of Kazak anti-plague research and production facilities [65].

Given the present international climate the apparent influence exerted by the Russian Federation over BW facilities in Kazakstan is probably not a matter of grave concern. Providing international assistance, dependent upon the ongoing dismantling of Kazakstan's military capability and the pursuit of civil programmes, continues to be forthcoming then it is likely that the infrastructure supporting the development and production of biological weapons will disappear. Such programmes have the added advantage of developing a Western management culture, allowing international monitoring of biopharmaceutical research programmes, and finally and perhaps crucially, of integrating Kazakstan's BW personnel within the international scientific community and preventing their exodus to countries pursuing the development of biological weapons.

---

43. Recent reports suggest that Mikhailichenko may no longer be Director of *Biomedpreparat* but is still closely involved with business ventures connected to the facility.

44. As an integral part of the contract which was to have been awarded for the sale of equipment to Stepnogorsk, *Giprobioprom* personnel were to receive extensive training from supplying biotechnology companies based throughout Europe and the USA.

8. References

1. Leitenberg, M. (1991), A return to Sverdlovsk: Allegations of Soviet activities related to biological weapons, *Arms Control*, **12**, pp. 161-190. Leitenberg, M. (1993), The biological weapons program of the former Soviet Union, *Biologicals*, **21**, 187-191. Leitenberg, M. (1994), The conversion of biological warfare research and development facilities to peaceful uses, in Geissler, E., Woodall, J.P. (eds.), *Control of dual-threat agents: The vaccines for peace programme*, Oxford University Press, Oxford, pp. 77-108; Leitenberg, M. (1996), Biological weapons and arms control, *Contemporary Security Policy*, **17**, 1-79. Rimmington, A. (1996), From military to industrial complex? The conversion of biological weapons' facilities in the Russian Federation, *Contemporary Security Policy*, **17**, 80-112.
2. Gertz, B. (5 April 1995), Germ warfare gives way to war on germs, *The Washington Times*.
3. Kazakh bacteriological center being dismantled by Americans (23 December 1997), Moscow NTV in Russian, 1100 GMT, 23 December 1997, translated in *FBIS*, FBIS-TEN-97-357.
4. *Kazakstan Defence Enterprise Directory* (August 1994), US Department of Commerce, Bureau of Export Administration. Interview with Gennadii Lepeshkin, General Director of the National Centre for Biotechnology of the Republic of Kazakstan, Science Industrial Complex *Biomedpreparat*, Stepnogorsk, 14 July 1995.
5. Hirsch, W. (undated), *Soviet BW and CW Preparations and Capabilities: Section II*, translated by Zaven Nalbandian and Ruth E. Anderson, Intelligence Branch, Plans, Training & Intelligence Division, Office of the Chief, Chemical Corps, Washington, D.C.
6. Zeberkhanuly, S. (23 May 1992), Aral disaster, Zhas alash, p. 2, translated in *JPRS*, JPRS-ULS-92-019, 2 July 1992, pp. 3-5. Starr, R.F. (1996), *The New Military in Russia: Ten Myths That Shape the Image*, Naval Institute Press, Annapolis, Maryland. Ponomarev, V. (1994), Biological and chemical weapons in Kazakhstan, *Ecostan News: Ecological News from Central Asia*, 2, No. 9, http://stsfac.mit.edu/projects/leep/Ecostan/Ecostan209.html, 12 January 1998.
7. Starr, R.F. (1996). Mayak Radio (Moscow 1200 gmt, 13 January 1992), translated in *Summary of World Broadcasts*, SU/1278, 15 January 1992, p. B/15. Zeberkhanuly, S. (23 May 1992), p. 2. Esirkepov, U. (16 May 1992), There is an island of revival in the Aral...: Who is declassifying the secrets, Molodezh uzbekistana, p. 5, translated in *JPRS*, JPRS-ULS-92-018, 23 June 1992, pp. 1-2. Leskov, S. (26 June 1993), Plague and the bomb: Russian and US military bacteriological programmes are being developed in deep secrecy, and present a terrible danger to the world, Izvestiya, p. 15, translated in *JPRS Report, Science & Technology, Central Eurasia: Life Sciences*, 27 August 1993, pp. 8-12. Krymova, S. (5-11 April 1991), ChP na ostrove vozrozhdeniya, *Poisk*, No. 15(101), p. 6.
8. King, J., Noble, J., Humphreys, A. (June 1996), *Central Asia*, Lonely Planet Publications, Hawthorn, Victoria.
9. Mamadaliev, S. (October 1996), The detection of probability of bringing in particularly dangerous infections to the territory of the Republic of Kazakstan and the development of scientifically based measures of control and prophylactics, in *Conversion in the Framework of International Collaboration, Proceedings*, Almaty, Kazakstan, pp. 46-47. Khailenko, N., Sedlovski, A., Roaliev, Sh. (October 1996), Alloplasmatic lines genetic and selectional peculiarities for spring wheat steady to phytopathogens, in *Conversion in the Framework of International Collaboration, Proceedings*, Almaty, Kazakstan, pp. 76-77.
10. Rimmington, A. (November 1995), *The Biopharmaceutical Industry in Kazakhstan: Opportunities for UK Companies*, Report of the DTI OSTEMS 'Scout' Mission to Kazakhstan, July 1995, Centre for Russian & East European Studies, The University of Birmingham.
11. Kosinov, A., Kotelnikova, O. (October 1996), Creation of the technology and man's gen-engineering insulin extraction (GEIE) on the basis of high productive GEIE producent, in *Conversion in the Framework of International Collaboration, Proceedings*, Almaty, Kazakstan, p. 11. Akhmatullina, N., Bugreev, V. (October 1996), Antimutagenesis as the basis of rehabilitation from the result of environmental pollution, in *Conversion in the Framework of International Collaboration, Proceedings*, Almaty, Kazakstan, pp. 94-95.
12. Tabakov, I.E. (1995), Dinamika prodazh produktov biotekhnologii na primere kormogo kontsentrata L-lizina kak otrazhenie makroekonimicheskoi situatsii v SNG, in *Perspektivy razvitiya proizvodstva biopreparatov dlya meditsiny I sel'skogo khozyaistva, Chast' 1, Razrabotka I sovershenstvovanie tekhnologii proizvodstva biopreparatov*, Stepnogorsk, p. 9. Gribalev, A.N. (1995), Ispol'zovanie fermentnykh preparatov v pishchevoi promyshlennosti, in *Perspektivy razvitiya proizvodstva biopreparatov dlya meditsiny I sel'skogo khozyaistva, Chast' 1, Razrabotka I sovershenstvovanie tekhnologii proizvodstva biopreparatov*, Stepnogorsk, p. 8.

13. *Zdravookhranenie Kazakhstana*, (April 1992), No. 4, p. 57. Interview with Alim Masgutovich Aikimbaev, Deputy Director, Kazak Scientific-Research Anti-Plague Institute, Almaty, 17 April 1995. *Kazakhstanskaya pravda* (16 April 1996). Anti-plague service created in Kazakstan (9 April 1992), Nezavisimaya gazeta, p. 6, translated in *JPRS Report, Science & Technology, Central Eurasia: Life Sciences*, JPRS-ULS-92-019, 2 July 1992, pp. 9-10.
14. Information provided by Vladimir Mikhailovich Stepanov, Director, Kazak Scientific-Research Anti-Plague Institute, Almaty, 17 April 1995.
15. Lepeshkin, G. (23 October 1996), Status and prospects for biotechnology development in the Republic of Kazakstan, ISTC Seminar on Conversion in the Framework of International Collaboration, Almaty, Kazakstan.
16. Cable to the US Secretary of State, 7 November 1996.
17. Sit'ko, N. (16-28 February 1995), Biotekhnologicheskii proryv ne za gorami.., *Nauka Kazakhstana*, No. 4(40), p. 1. *Kazakhstanskaya pravda* (8 February 1996).
18. Malleret, T. (1992), *Conversion of the defence industry in the former Soviet Union*, Institute for East-West Security Studies, Occasional Paper No. 23, Westview Press, New York. Cooper, J.M. (1988), Civilian and military engineering, in Berry, M.J., *Science & Technology in the USSR*, Longman, Harlow.
19. Koishibaev, M.S. (1994), Na krutom povorote vremeni, in *Stepnogorsk. Zdes' propisany nashi serdtsa. stranitsy istorii goroda Stepnogorska Akmolinskoi oblasti*, Atamura, Almaty, p. 1.
20. Perov, E.I., Trudnye shagi "Progressa" (1994), in *Stepnogorsk. Zdes' propisany nashi serdtsa. stranitsy istorii goroda Stepnogorska Akmolinskoi oblasti*, Atamura, Almaty, pp. 107-125.
21. Lucke, E-M., Poetzsch, E. (Eds.), *Biotechnology Directory Eastern Europe*, Walter de Gruyter, Berlin & New York, 1993. Interview with Yurii Ivanovich Kondrashin and Sergei L. Osipovsky at Science Production Association *Biomash*, Moscow, 31 May 1993. Ivchenko, L. (28 January 1993), A private firm has privatised a major medical factory for a song, Trud, 28 January 1993, translated in *JPRS Report, Science & Technology, Central Eurasia: Life Sciences*, pp. 22-24.
22. Lepeshkin, G.N. (1994), Flagmanu farmindustrii - byt', in *Stepnogorsk. Zdes' propisany nashi serdtsa. Stranitsy istorii goroda Stepnogorska Akmolinskoi oblasti*, Atamura, Almaty, pp.126-132.
23. Korotko ob avtorakh (1994), in *Stepnogorsk. Zdes' propisany nashi serdtsa. Stranitsy istorii goroda Stepnogorska Akmolinskoi oblasti*, Atamura, Almaty, p. 202.
24. Interview with Gennadii Lepeshkin, General Director of the National Centre for Biotechnology of the Republic of Kazakhstan, Science Industrial Complex *Biomedpreparat*, Stepnogorsk, 14 July 1995. Nesterenko, S., Kriger, V. (28 May 1993), Ot supersekretnykh vaktsin k sredstvu bor'by so SPIDom. Takov potentsial Stepnogorskogo kompleksa "Biomedpreparat", *Kazakhstanskaya pravda*, No. 122, p. 7.
25. Domaradskii, I. (December 1996), Istoriya odnoi avantyury, *Znanie-sila*, No. 11(833), p. 63.
26. Highfield, R. (21 May 1991), Soviet drugs could be used to treat UK patients, *Daily Telegraph*, p. 4. *NPO Vector, Novosibirsk, Siberia, USSR, Biological Sciences - Research and Production, Products* (undated), distributed by Oxford Virology plc, London. Telex from Professor Alibekov to Derek Lennon, Chairman of Oxford Virology plc, 12:46, 5 December 1990. Copy kindly supplied by Mr Lennon.
27. Adams, J. (1994), *The New Spies: Exploring the Frontiers of Espionage*, Hutchinson, London. Urban, M. (1996), *UK Eyes Alpha: The Inside Story of British Intelligence*, Faber and Faber, London, 1996.
28. O merakh po obespecheniyu deyatel'nosti Natsional'nogo tsentra po biotekhnologii Respubliki Kazakhstan (16 November 1993), *Postanovlenie kabineta ministrov respubliki Kazakstan*, No. 1140. Sozdan Natsional'nyi tsentr po biotekhnologii (1-15 January 1994), *Nauka Kazakhstana*, No. 1(13), p. 1. National Centre on Biotechnology Republic of Kazakhstan, document provided to Dr Anthony Rimmington by Yurii Aleksandrovich Shapovalov, Deputy Director of the National Biotechnology Centre, Institute of Experimental Biology, Almaty, 21 April 1995.
29. Kazakh Radio (0500 gmt, 1 March 1992), Almaty, translated in *Summary of World Broadcasts*, SU/W0221, 13 March 1992, p. A/7. Pavlov, Yu. (13 February 1992), Kazakhstan has its own pharmaceutical concern, Kazakhstanskaya pravda, p. 4, translated in *JPRS Report, Science & Technology, Central Eurasia: Life Sciences*, JPRS-ULS-92-015, 27 May 1992, pp. 13-14.
30. Gil'manov, M. (16-28 February 1994), Gorizonty biotekhnologii, *Nauka Kazakhstana*, No. 4(16), pp. 1-2.
31. Interview with Yurii Aleksandrovich Shapovalov and Roza Ultybaevichi Beissimbaeva, Branch of the National Biotechnology Centre, Almaty, 21 April 1995.
32. Sit'ko, N. (16-28 February 1995), Biotekhnologicheskii proryv ne za gorami?.., *Nauka Kazakhstana*, No.4(40), p. 1.
33. Molchanova, A. (5 April 1994), Lekarstv stanet bol'she, no ne zavtra, *Kazakstanskaya pravda*, No. 47(21,543), p. 2.

34. Daleev, M. (11 August 1995), Kazakstanu nuzhna i farmatsevticheskaya nezavisimost', *Kazakstanskaya pravda*, No. 154(21859), p. 3.
35. Decree of the board of the Ministry of Health (April 1992), Zdravookhranenie Kazakhstana, No. 4, p. 57, translated in *JPRS Report, Science & Technology, Central Eurasia: Life Sciences*, JPRS-UST-94-027, 31 October 1994, pp. 41-42.
36. *Kazakhstanskaya pravda* (16 April 1996).
37. Ganzha, V. (7 April 1992), Unified Antiplague Service Formed in Kazakstan, TASS, carried on *FT Profile*, 11:04:06, 14 October 1997.
38. Bakhmat, G.D. (1995), AO "Progress" - flagman mikrobiologicheskoi promyshlennosti Kazakhstana, in *Perspektivy razvitiya proizvodstva biopreparatov dlya meditsiny I sel'skogo khozyaistva, Chast' 1, Razrabotka I sovershenstvovanie tekhnologii proizvodstva biopreparatov*, Stepnogorsk, p. 7.
39. Upcoming sales, company database, http://www.kazecon.kz/StockScripts/getinfo.idc?kod=2409, 18 November 1997.
40. Decree of the Government of the Republic of Kazakstan "The Issues of State Property Privitisation", No. 499 (23 April 1996), http://www.kazecon.kz/StockInfo/English/individ2.htm, 12 January 1998.
41. Russia's Radio (1600 gmt, 9 May 1992), translated in *Summary of World Broadcasts*, SU/W0231, 22 May 1992, p. A/11.
42. Smith, R.J. (1 September 1992), US fears Moscow still makes germ weapons, *International Herald Tribune*, pp. 1, 6.
43. Ellis, W.S. (February 1990), The Aral: A Soviet sea lies dying, *National Geographic*, **177**, No. 2, pp. 73-92.
44. Neiman, B.Ya. (1983), *Industriya mikrobov*, Znanie, Moscow. Rychkov, R.S. (1981), *Mikrobiologicheskaya promyshlennost' - meditsine, Fundamental'nye -meditsine*, Nauka, Moscow.
45. Interview with Gennadii Lepeshkin (Director of the National Centre for Biotechnology of Kazakhstan) and Aleksandr Kosinov (Director of the Institute of Pharmaceutical Biotechnology), Moscow, 17 August 1995.
46. Leninskaya smena (4 April 1991), p. 2, translated in *JPRS*, ULS-91-016, 10 September 1991, pp. 31-32.
47. Nesterenko, S., Kriger, V. (28 May 1993), Ot supersekretnykh vaktsin k sredstvu bor'by so SPIDom. Takov potentsial Stepnogorskogo kompleksa "Biomedpreparat", Kazakhstanskaya pravda, No. 122, p. 7.
48. Interview with Aleksandr Petrovich Mikhailichenko (Director *KNPK Biomedpreparat*), Stepnogorsk, 13 July 1995.
49. Interview with UK distributer of medical and biopharmaceutical equipment, 21 October 1996.
50. Moscow Radiostantsiya Ekho Moskvy (0800 GMT, 15 September 1995), translated in *FBIS Report, Arms Control & Proliferation Issues, Central Eurasia, Russia*, FBIS-TAC-95-005, 15 September 1995.
51. The Asian Age (9 July 1995), pp. 1, 4, carried in *FBIS Report, Environment & World Health, World Health, India*, FBIS-TEN-95-012, 7 September 1995, http://wnc.fedworld.gov/cgi-bin/r...dm9q&CID=C93499755859375077510907, 10:05:15, 22 October 1997.
52. Meeting with representative of UK equipment distributor, Moscow, 4 June 1993. *Akmola entsiklopediya* (1995), "Atamura", Almaty. *Biznes-karta 97, Meditsina, Proizvoditeli medikamentov, vetpreparatov, medtekhniki, Strany SNG (vklyuchaya Rossiyu) I Baltii* (1997), Agenstvo delovoi informatsii "Biznes-karta", Moscow. Yuferev, O.V. (1993), *Biznes-karta 93, Meditsinskaya promyshlennost'*, Agenstvo Delovoi Informatsii, Moscow. *Biznes-karta 94, Meditsinskaya promyshlennost'* (1994), "Biznes-karta", Moscow.
53. Interview with Gabdulla Gabdullovich Buzurbaev (Director, Medico-Biological Institute), Stepnogorsk, 14 July 1995.
54. Interview with Nikolai Stefanovich Shevyrev (Director), Vyacheslav Mikhailovich Bezgin (Chief Engineer) and Vladimir Egorovich Kozlov (Head of Central Research Laboratory), Main Administration Building, Kursk Biofactory, Kursk, 16 August 1994.
55. Cooper, J.E. (September 1996), Links with Kazakhstan, *Biologist*, *43*, No. 4.
56. *Mezhdunarodnaya nauchno-prakticheskaya konferentsiya: Perspektivy razvitiya proizvodstva biopreparatov dlya meditsiny I sel'skogo khozyaistva* (1995), Stepnogorsk.
57. Giebink, D., International Science & Technology Centre activity on development and implementation of conversion R & D projects in Kazakstan and other CIS countries, ISTC Seminar Conversion in the Framework of International Collaboration, 23 October 1996, Almaty, Kazakstan.
58. *Governing Board Statement. Projects Approved for Funding at the Tenth Meeting of the ISTC Governing Board* (26 June 1996), http://www.istc.ru/gbs10a.htm, 12 January 1998. *The International Science and Technology Center Annual Report* (January-December 1996), Moscow.

59. *The Nuclear Roundtable Background Documents, The Co-operative Threat Reduction Assistance to Kazakstan: Co-operative Threat Reduction Program, Department of Defence* (January 16, 1997), http://www.stimson.org/rd-table/ctr-kaz.htm, 12 January 1998.
60. Ssha podderzhat konversiyu v Kazakhstane (13 August 1994), *Kazakhstanskaya pravda*, No. 98(21,594), p. 1. Ardaev, V. (8 April 1995), It is better to sell vitamins and not bacteriological weapons, *Izvestiya*. Perry visits Almaty (6 April 1995), *OMRI Daily Digest*, No. 69, Part 1.
61. 18 Jul 95 International Market Insight Reports: Kazakhstan: Almaty Business Roundtable (April 7 '95), Asia Intelligence Wire, *FT Profile*, 10:35, 14 October 1997.
62. 16 May 96 ICN Pharma, Wash DC Firm in Kazakh Joint Venture, Dow Jones News Database, *FT Profile*, 10:33, 14 October 1997; Mar 97, ICN Expands its Global Business through Acquisitions, Predicasts PROMT, Med Ad News, *FT Profile*, 10:30, 14 October 1997; 01 Jun 96, ICN Forms Joint Venture, Predicasts PROMT, Applied Genetics News, *FT Profile*, 10:32, 14 October 1997; 15 May 96 US Drug Company to Convert Kazakh Bio-Weapons Plant, The Associated press, *FT Profile*, 10:33, 14 October 1997; 14 May 96 ICN Pharmaceuticals and Allen & Associates to Convert Former Soviet Scientific Production Complex into.., Reuter Textline, PR Newswire, *FT Profile*, 10:34, 14 October 1997.
63. Akiner, S. (1996), Potential threats to stability and social cohesion in Central Asia, in Weichhardt, R. (Ed.), *Economic Developments and Reforms in Cooperation Partner Countries: The Social and Human Dimension*, Colloquium 26-28 June, 1996, NATO Economics Directorate, Brussels, pp. 47-60.
64. Sandakhchiev, L.S. (1991), Innovation development program of the Association "VECTOR" for 1992-1995 and till 2005, Novosibirsk.
65. Chtoby chuma na nashu golovu ne svalilas' (23 May 1997), *Meditsinskaya gazeta*, No. 39(5760), p. 1.

FACILITATING THE CONVERSION OF BTW FACILITIES THROUGH INTERNATIONAL TECHNICAL COOPERATION: BTWC IMPLICATIONS

IRIS HUNGER
Max Delbrück Centrum for Molecular Medicine
Berlin-Buch, Germany

1. Introduction

Conversion of biological and toxin weapons (BTW) facilities is a challenging, increasingly important topic although the number of states that have to convert such facilities is limited. Relevant international treaties give very little guidance in regard to conversion. The existing limited conversion efforts have so far been undertaken in a bilateral or national framework. [1]

Two simple observations underline, however, the importance of pursuing conversion efforts in an international cooperative setting. First, the conversion of BTW facilities is of international interest. That is because this difficult, highly sensible undertaking has not only consequences for the state concerned or for the respective facility and its employees. The issue of whether any given BTW facility is converted or not influences the security of nations world-wide. Second, conversion depends on international cooperation. Respective facilities will have to provide openness and transparency. Conversion will further necessitate exchange opportunities and training and education programmes for the facility's scientists as well as technology transfer to make use of existing knowledge, experiences and equipment in a way that meets international standards. Multilateral financing arrangements might also be necessary. It has been argued that efforts to foster international technical cooperation in the framework of Article X of the Biological and Toxin Weapons Convention (BTWC) might contribute at the same time to the conversion of BTW facilities. [2] This paper intends to examine the relationship between international technical cooperation under the BTWC and conversion of BTW facilities.

I will first outline the technical cooperation aspects of the BTWC, the BTWC being the most important and comprehensive tool to control BTW. I will furthermore describe the work done on technical cooperation in the Ad Hoc Group (AHG) of States Parties to the BTWC that is on its way to developing a Protocol for promoting compliance with the Convention. I will then take a look at the different ideas of how to implement the technical cooperation requirement under the BTWC and ask whether these programmes are able to support conversion. Finally I will draw conclusions and propose a model programme for conversion of BTW facilities. I will argue that

E. Geissler et al. (eds.), Conversion of Former BTW Facilities, 187-199.
© 1998 *Kluwer Academic Publishers. Printed in the Netherlands.*

conversion programmes should be kept distinctly separate from any efforts to implement the technical cooperation requirement under the BTWC for different reasons, but that special aspects and ideas of programmes for technical cooperation can be adopted for the establishment of conversion programmes.

2. The Technical Cooperation Requirement in the Biological and Toxin Weapons Convention

The article of the BTWC that concerns technical cooperation and assistance is Article X, that reads as follows:

"(1)The States Parties to this Convention undertake to facilitate, and have the right to participate in, the fullest possible exchange of equipment, materials and scientific and technological information for the use of bacteriological (biological) agents and toxins for peaceful purposes. Parties to the Convention in a position to do so shall also cooperate in contributing individually or together with other States or international organizations to the further development and application of scientific discoveries in the field of bacteriology (biology) for prevention of disease, or for other peaceful purposes.

(2)This Convention shall be implemented in a manner designed to avoid hampering the economic or technological development of States Parties to the Convention or international cooperation in the field of peaceful bacteriological (biological) activities, including the international exchange of bacteriological (biological) agents and toxins and equipment for the processing, use or production of bacteriological (biological) agents and toxins for peaceful purposes in accordance with the provisions of the Convention." [3]

Article X comprises two aspects. The first is the promotional aspect, demanding the facilitation of and codifying the right of participation in the exchange of relevant equipment, materials and knowledge for peaceful purposes. The second is the regulatory aspect, which forbids, while implementing the Convention, the hampering of economic or technological development. The two aspects are considerably different. Concerning the regulatory aspect, expressed in the second part of Article X, an especially close relationship exists with Article III of the BTWC that calls on States Parties not to assist in any way other states, groups of states or international organizations in acquiring BTW. This paper will not deal with that line of discussion but will concentrate on the obligations to contribute to and the right of participating in the fullest possible exchange of equipment, materials and scientific and technological information for the use of bacteriological (biological) agents and toxins for peaceful purposes.

The aspect of technical cooperation in the framework of the BTWC had not played an important role until the beginning of the 1990s. In the emerging discussions about improving the Convention's efficiency the topic of implementing Article X has become more and more important. It was especially taken up by states of the Non-Aligned Movement, like Brazil, India and Iran, and by a few non-governmental organizations (NGOs). This growing interest is driven by several factors. First, the BTWC is on the threshold of being developed into a more effective arms control treaty, certainly involving different kinds of more or less burdensome controls and restrictions for the biotechnological industry. Second, biotechnologies promise to be one of the most important technologies in the future able to facilitate many solutions to the especially pressing problems in the fields of health care, environmental protection and nutrition. And third, the technology gap between developed and developing countries is still rapidly growing. Accordingly, when the mandate of the AHG, that was established in 1994 to draft proposals for a legally binding instrument, was accepted, the implementation of Article X was included as one of four important points of consideration as follows:

"The Ad Hoc Group shall, inter alia consider:

- Definitions of terms and objective criteria, [...];
- The incorporation of existing and further enhanced confidence building and transparency measures, [...];
- A system of measures to promote compliance with the Convention, [...];
- Specific measures designed to ensure effective and full implementation of Article X, which also avoid any restrictions incompatible with the obligations undertaken under the Convention, noting that the provisions of the Convention should not be used to impose restrictions and/or limitations on the transfer for purposes consistent with the objectives and the provisions of the Convention of scientific knowledge, technology, equipment and materials." [4]

The implementation of Article X and the subsequent comprehensive fulfilment of the AHG mandate have recently become more and more the subject of debate. The range of disagreement was nicely illustrated during the Fourth Review Conference in 1996. During the General Debate the majority of delegations took up the topic of Article X implementation in their statements. Western countries stressed the cooperative arrangements that are already in force in the biological field. Countries of the Non-Aligned Movement called in their statements for much greater efforts to implement Article X and not to attach greater importance to some articles of the Convention than to others. Bangladesh pointed out that otherwise "many in the developing world would see little merit in the ongoing exercise" [5] of strengthening the Convention. It was mentioned several times that without strong provisions for technical cooperation the universal adherence to the BTWC would be difficult to obtain. Surprisingly few states referred to the work done in the AHG concerning the implementation of Article X. It was probably assumed that, whereas the substantial work is done in the AHG, the Review Conference was a place to voice principle beliefs one more time.

3. The Discussions on Technical Cooperation in the Ad Hoc Group

Since the first substantial session of the AHG in July 1995, considerably less attention, at least in terms of time allocated, has been given to Article X measures than to other aspects of the mandate. Only 15% of the meetings during the second to eighth session of the AHG were devoted to this topic (see Table 1). Presumably this reflects the fact that compliance measures, i.e. declarations and various types of visits and investigations, and the respective definitions of terms and objective criteria are seen as the core elements of the future Protocol. Considering the obvious need for sound Article X mechanisms and the existing difficulties to agree on how to implement Article X, this failure so far to address comprehensively Article X and its effective and full implementation is, nevertheless, alarming.

Until March 1997 discussions in the AHG took place in a more general way, aided by Friends of the Chair. Under the guidance of the Friend of the Chair on Article X several working papers had been prepared outlining among other things "Elements for Structured Discussions on Article X" and "Activities of Multilateral Cooperation in Areas related to the BWC and their Relevance for Cooperation under Article X of the BWC". [6] It was encouraging to note that emphasis had been placed on the need to focus on a specific range of activities such as activities in "areas directly relevant to the Convention". However, it is disturbing that such focusing has taken place only to a very limited extent even until now. [7]

Table 1. Number of meetings on different topics during different sessions of the Ad Hoc Group

	Number of meetings concerning compliance measures	Number of meetings concerning definitions	Number of meetings concerning Article X measures	Number of meetings concerning confidence-building and transparency measures
Second session (Jul 1995)	9	7	2	2
Third session (Nov/Dec 1995)	6	5	4	2
Fourth session (Jul 1996)	7	6	3	2
Fifth session (Sep 1996)	7	4	4	2
Sixth session (Mar 1997)	8	6	5	0
Seventh session (Jul/Aug 1997)	11	5	3	0
Eighth session (Sep/Oct 1997)	16	3	3	0
Sum (% of all 163 meetings) [8]	64 (39%)	36 (22%)	24 (15%)	8 (5%)

At the seventh session in July/August 1997 a rolling text of the future Protocol was introduced which was discussed and further developed during that session and the eighth session in September as well. This rolling text contains Article VII on "Scientific and technological exchange for peaceful purposes and technical cooperation", which describes in general terms the activities which States Parties shall implement. [9] These activities include among others the following:

-promoting the publication of information about current research programmes of relevance to the Convention;

-promoting the establishment of research institutes for the examination of biological agents and toxins;
-supporting the establishment of relevant biological electronical data bases; and
-promoting the transfer of relevant technology for peaceful uses.

It has also been suggested to promote expert training in BTW defence activities, and to make equipment and technology available and to promote collaborative research and development for BTW defence. These suggestions are presumably an attempt to focus Article X measures on topics of relevance to the BTWC. Undoubtedly, BTW defence is a highly relevant area for measures under Article X. However, multiple square brackets, which are used around the respective phrases in cases where states are not yet able to agree on the content or on specific formulations, illustrate the large degree of disagreement as to whether such activities should be pursued cooperatively and consequently under complete international transparency.

Article VII of the rolling text also describes some institutional structures that the future BTWC Organization is envisaged to set up on a cooperative basis, including mechanisms for the establishment and functioning of necessary national authorities, for the fulfilling of declaration requirements, for the drawing up of necessary internal legislation, and for the inclusion of a cooperative dimension in non-challenge visits.

As for other articles of the Protocol there is also an annex for Article VII. Annexes describe the technical details of the implementation of respective articles. Unfortunately, under Annex F on "Scientific and technological exchange for peaceful purposes and technical cooperation" one finds only an empty page. No progress has been made so far in the AHG to fill this gap.

4. Proposed Programmes for the Implementation of Article X of the BTWC

Different ideas exist as to how Article X of the BTWC should be practically implemented. Industrialized states are convinced that Article X is already fully implemented by means of existing cooperation agreements in areas such as health care, agriculture and environmental protection. Numerous states of the Non-Aligned Movement claim that comprehensive implementation of Article X has still to take place. Four of the programmes and organizations that are more often mentioned as possible mechanisms for furthering implementation of Article X are shortly described in the following.

4.1. WORLD HEALTH ORGANIZATION (WHO)

The WHO is very often mentioned as an international organization that could at least support the implementation of Article X. It has even been argued that the WHO could assume responsibility for Article X implementation. Of special relevance to the BTWC is the WHO's Division of Emerging and other Communicable Diseases Surveillance and Control (EMC). This division envisages an alert world that is able to contain communicable diseases through strong national disease surveillance and control programmes, global networks of centres, organizations and individuals to monitor diseases, rapid information exchange through electronic links to guide policies, international collaboration, trade and travel, as well as effective national and international preparedness and rapid response to contain epidemics of international importance.

The total budget for this division was US$ 16.4 million for the two-year period 1996-1997. Activities include dissemination of information, reviewing international health regulations, assessing, promoting and strengthening disease surveillance and control systems, operational research on such systems, epidemic preparedness and response, monitoring of drug resistance of relevant micro-organisms, providing training and training material in immunology, vaccinology, biotechnology and biosafety as applied to communicable diseases, and research on field diagnostic techniques. Some of the diseases dealt with in this division are of clear relevance for the BTWC, like plague, viral encephalitis and yellow fever. Others, like influenza and AIDS, are of limited importance under the BTWC.

The EMC is collaborating with the division of emergency and humanitarian action to improve the WHO's capacity to respond to outbreaks of infectious diseases which represent major public health threats. This capacity represents a significant advantage also in the case of large-scale military or terrorist use of BTW. However, the WHO's focus is on international health issues and not on BTW control. Accordingly, the WHO can, out of its own structure, provide support on certain issues of interest to the BTWC, but it is not able to provide a specific mechanism for the implementation of Article X. Moreover, WHO has shown some reluctance, at least until recently, to become involved in the political business of BTW control.

4.2. PROGRAMME FOR MONITORING EMERGING DISEASES (ProMED)

This programme calls for a network of about 10 medical centres, strategically located in the developing world, acting as global watchdogs which look for outbreaks of new emerging diseases. These centres would be chosen from among existing facilities on the basis of their capabilities, and would receive priority for international assistance. The centres would be linked electronically and through the exchange of personnel and would share expertise and information. Early warning information of outbreaks of known and unknown diseases, detected and analysed by the watchdog centres, would be rapidly transmitted to the WHO in order to mobilise international assistance in a timely manner so as to prevent epidemic spread. A publicly accessible system on the Internet, called ProMED-mail, has been operating since 1994, reporting on outbreaks of emerging diseases around the world and moderating discussion on them. Its e-mail list reaches over 10,000 subscribers in more than 130 countries, and its website reaches thousands more. ProMED-mail has occasionally reported new outbreaks faster than the WHO, and has been instrumental in bringing aid to health authorities dealing with, e.g., dengue in India and tuberculosis in Romania, and to private citizens with rare diseases in China and elsewhere.

Although the idea of ProMED was originally developed in order to support the implementation of Article X of the BTWC, this programme's focus is now world health promotion by monitoring diseases in general. It neither deals exclusively with agents relevant to the BTWC, nor is participation restricted to States Parties. Therefore, it is not able to provide a specific mechanism for the comprehensive implementation of Article X.

4.3. INTERNATIONAL CENTRE FOR GENETIC ENGINEERING AND BIOTECHNOLOGY (ICGEB)

This independent international organization was established as a joint project with UNIDO ten years ago and has currently around 40 member states, mostly from developing countries. Its aim is to provide access to advanced technology in the developing world by providing research, training and institutional support in the field of biotechnologies. The activities are mostly directed towards medical and agricultural problems with special focus on the needs of the member states. The ICGEB comprises two research components, one in Trieste/Italy and one in New Delhi/India, and affiliated centres. The scientific population includes more than 200 people from 28 countries. Besides the research activities in the two centres the ICGEB supports collaborative research programmes in the affiliated centres, provides extensive short- and long-term training courses and fellowships and maintains the ICGEBnet, a computer resource for biotechnology.

The ICGEB has often been proposed as an organization that could be responsible for the implementation of Article X. In its 1996 Activity Report the Director of the ICGEB is urged to "pursue his efforts aimed at providing the States Parties to the Biological Weapons Convention with the Centre's expertise on all issues related to technical assistance, training and technology transfer, with a view to the peaceful use of biotechnology foreseen by Article 10 of the Convention". [10] However, the activities of the ICGEB do not specifically focus on relevant BTWC areas and are therefore only to a limited extent suitable as mechanisms for the implementation of Article X.

4.4. PROGRAMME FOR COUNTERING EMERGING INFECTIOUS DISEASES (ProCEID)

The Programme for Countering Emerging Infectious Diseases by prophylactic, diagnostic and therapeutic measures (ProCEID), is a proposal specifically designed for the purpose of strengthening the BTWC by implementing Article X. This proposal is the result of numerous discussions of an earlier proposal, the Vaccines for Peace Programme. [11]

Key objectives of ProCEID include, *inter alia*, an increase of world capacity to produce and make available prophylactic, diagnostic and therapeutic measures for pathogens of relevance to the BTWC, thereby preventing and combatting the respective diseases, enhancing peaceful international cooperation in molecular medicine and biotechnology and strengthening the BTWC by implementing Article X. In contrast to previous programmes, ProCEID would involve the development and production of all kinds of biologicals necessary to counter diseases of relevance to the BTWC. Nevertheless, research on and development and production of vaccines would be an outstanding topic of ProCEID not only because of global health demands but also because of the outstanding role vaccines play in defensive and offensive military biological activities.

How would this programme work? ProCEID would be executed by international teams in institutions located in different parts of the world under conditions of complete transparency. The facilities involved would be provided with the equipment, know-how and agents necessary to develop and produce prophylactic, diagnostic and therapeutic measures which are essential for the protection against relevant infectious diseases. These measures would be provided on request to States Parties for countering the military or terrorist use of BTW agents as well as natural outbreaks of diseases.

Most of the activities involved in such a programme are of a dual-use nature. To minimise the risks of misuse of the programme, ProCEID would only be open to States Parties in good standing with the BTWC, that is, to states that have ratified the future Protocol to the Convention, and that are also States Parties to the Chemical Weapons Convention. Extreme care would have to be taken that participation in ProCEID would not interfere with the BTWC obligation to avoid proliferation of weapons-related agents, technologies and knowledge. This can be ensured by a rigorous transparency regime. In the long run, ProCEID can even contribute to the international control of the spread of relevant agents and technologies, a spread that will definitely take place because of ongoing biotechnological development worldwide.

Alleviation of the suspicion created by research on, and development and production of biologicals protecting from BTW relevant diseases and the concurrent provision of these biologicals --which so far have only marginally been covered by industry and international health care programmes-- without weakening (national) military defence activities represent outstanding global advantages of ProCEID. This programme would promote the establishment of an international system of biological security and fulfil the obligations of Article X of the BTWC.

Additional benefits of ProCEID may include "contributing to conversion of former BW facilities to peaceful purposes under international control". [12] It is reasoned that using former BTW facilities would not only contribute significantly to reducing costs, time, personnel, and numerous technical prerequisites necessary to establish ProCEID, but would for the first time contribute to establishing an efficient conversion programme while minimising the dangers of misuse and clandestine reversion involved in such undertakings.

As clear as the key objectives and methods of operation are, a number of problems and questions remain to be dealt with before this programme could actually be put into action.

5. Conversion Aspects

The BTWC itself covers the topic of conversion only to a limited extent. Under its Article II States Parties are requested to destroy or divert to peaceful purposes agents, toxins, weapons, equipment and means of delivery. The conversion of the respective BTW facilities is not mentioned.

> "Each State Party to this Convention undertakes to destroy, or to divert to peaceful purposes, as soon as possible but not later than nine month after the entry into force of the Convention, all agents, toxins, weapons, equipment and means of delivery specified in article I of the Convention, which are in its possession or under its jurisdiction or control. In implementing the provisions of this article all necessary safety precautions shall be observed to protect populations and the environment." [13]

This is in striking contrast to the Chemical Weapons Convention where conversion is explicitly regulated. The conversion of chemical weapons production facilities is only allowed either for the purpose of destroying old chemical weapons for a limited period or on special request. Such conversion programmes need the agreement of the Technical Secretariat or the Conference of States Parties respectively and fall under rigid declaration and verification provisions.

Moreover, the Ad Hoc Group, working on the development of a comprehensive Protocol to ensure compliance with the BTWC, has not yet dealt at all with the problem of conversion. Conversion is neither mentioned as a concern per se nor is it addressed in connection with discussions on technical cooperation under the BTWC. Programmes that have been proposed for the implementation of Article X do not touch on the topic of conversion, with the exception of ProCEID. When the AHG analysed the relationship between Article X and other articles of the Convention, it did not identify a connection between Articles II and X. [14]

Why is it that no discussion on conversion takes place in the AHG? And why is it, that no connection has been established between discussions on programmes for technical cooperation under the BTWC and conversion programmes? Obviously - conversion of BTW facilities is a concern that applies only to a very limited number of states. Most of the relevant facilities that were located in states which are now States Parties to the BTWC ceased to function years ago, before or at the time when the BTWC entered into force. Talking about conversion would require talking about past BTW programmes, a task that obviously does not come easy to some states.

The explanation for the lack of connection between discussions on programmes for technical cooperation under the BTWC and conversion programmes lies in some fundamental differences between such programmes. The basic difference is that one process - conversion - is the result of the desire to stop activities that are prohibited by the BTWC, whereas the other - technical cooperation - results from the desire to provide specific incentives to stick to the provisions of the BTWC in order to prevent, from the very beginning, such prohibited activities from happening. Politically, it thus seems to send the wrong signal to use facilities that have been involved in activities which are prohibited by the BTWC for activities intended to provide incentives to stick to the provisions of this Convention.

Besides this basic difference there are other reasons why conversion programmes remain separated from programmes for technical cooperation under the BTWC and its future Protocol.

1. Conversion programmes are only relevant for a very limited number of states, whereas programmes for technical cooperation should involve a maximal number of States Parties in order to fulfil the function of providing incentives for joining the BTWC and its future Protocol and sticking to their provisions.

2. Conversion programmes are relevant for states independent of their status in respect to the BTWC, whereas programmes for technical cooperation under the BTWC are available for States Parties only. With the future existence of a Protocol to the BTWC, which will contain provisions for the implementation of Article X, such programmes will be available for states only that are States Parties both to the BTWC and its Protocol.

3. Conversion programmes take place in states that have at least modest, if not very high capabilities in biotechnology, whereas programmes for technical cooperation are directed primarily towards states that have either very limited or no capabilities at all in this regard. Programmes for technical cooperation focus on countries with limited biotechnological capabilities in order to attract them to the treaty and its future Protocol --even in cases where BTW control is of limited relevance and respective financial resources are scarce.

Table 2. Differences between conversion programmes and programmes for technical cooperation under the BTWC

	Conversion programmes	Programmes for technical cooperation under the BTWC
Number of states concerned	small	high
Status of the states concerned in respect to the BTWC	member states, signatories, other countries	member states
Existence of a bio-technological industry in recipient states	existent	not always existent
Time frame	limited	open ended
Source of financing	state(s) concerned and a limited number of supporter states	all member states of the BTWC
Degree of suspicion	high	low

4. Conversion programmes will take place only for a limited time period, until conversion is completed, whereas programmes for technical cooperation do not have immanent limitations because the technology gap between developed and developing countries may persist for quite a long time.

5. Conversion programmes are financed by the state or states concerned, possibly supported by a limited number of additional concerned states, whereas programmes for technical cooperation in the framework of the BTWC will be financed by all

States Parties. There is no justification for all States Parties to pay for the conversion of facilities in single states that had been involved in activities which are prohibited by the BTWC.

6. Finally, BTW facilities are by their very nature highly suspicious, although conversion is aimed at resolving such suspicions. Facilities where programmes for technical cooperation are envisaged will always be under special control because of their dual-use activities and their work, at least in part, on dual-threat agents. There is thus no point in multiplying the degree of suspicion by using highly suspicious BTW facilities to perform sensitive cooperation activities.

These differences precisely lead to the conclusion that conversion programmes should be pursued separately from any programmes aimed at implementing the technical cooperation requirement under Article X of the BTWC. Should the AHG take up the topic of conversion, it should not mix its aim to strengthen Article II of the Convention with discussions on technical cooperation under Article X. This is not to say that programmes for technical cooperation could not be performed in BTW facilities that are in the process of conversion so as to support that conversion process. At least for political reasons, however, such connections should be rare exceptions.

Several aspects of programmes for technical cooperation are also relevant for and may support the development of conversion programmes. For example, both types of programmes need mechanisms to minimize the risk of misuse. Such risk minimization can be achieved through a rigid system of transparency and openness including provisions for the exchange of personnel and for joint projects with other --civilian-- facilities as well as by minimizing restrictions placed on investigation teams. Another common aspect is economic viability. Both types of programmes must include prospects for becoming economically independent in the foreseeable future. This requires analysing the existing (pharmaceutical) market, identifying gaps and subsequently opportunities for marketable products, and balancing these against the capabilities of the facility.

6. Conclusion and outlook

Although conversion of BTW facilities requires technical cooperation, we must not rely on the results of discussions on technical cooperation with respect to implementing Article X of the BTWC. In addition, each topic should be dealt with independently for reasons outlined above. Indeed, specific conversion programmes must be developed for every single BTW facility. The basic prerequisite for such an undertaking is the sound political decision of key political leaders to stop BTW activities. This decision should subsequently be accompanied by a detailed conversion plan specifying, *inter alia*, time frames, types of activities, the structure of working groups, financing patterns, control measures, transparency mechanisms etc. Based upon proposals for programmes for technical cooperation and results of discussions on transparency mechanisms, a basic outline for a conversion programme is, finally, given in the following:

1. *Each individual facility works on a topic of economic viability and makes maximal use of existing equipment and expertise.* As mentioned before, such an approach requires analysing the market situation and the capabilities of the facility to be converted. It will usually be easier to focus on diagnostics than on prophylactic or therapeutic applications to humans, because safety regulations are not as rigorous. In special circumstances it may also be possible to produce certain pharmaceutical products for a defined, limited market, like vaccines for the local market, that do not require the complete range of safety provisions that is necessary for the international market.

2. *Top leading positions are filled with civilian trained international experts, excluding former leaders of the facility.* Usually BTW facilities are under military control and have a highly hierarchic decision-making structure. Reassignment of leading positions is therefore an effective way to change the activities of a facility. This approach may involve the dismissal of people with highly relevant knowledge of both the facility and the whole structure of the former BTW programme. However, ensuring a sound change of activities of a BTW facility, including the decision-making structure and the work culture, is greatly facilitated by substituting people in leading positions.

3. *Working groups consist of employees of the facility and international civilian experts.* This approach fosters the development of a new work culture, facilitates the introduction of new scientific and administrative techniques and knowledge, promotes the international dissemination of expertise and generally supports the integration of the employees of the BTW facility into the international scientific community, thereby increasing openness.

4. *The facility provides continued access to investigation teams.* These investigations are carried out according to the mechanisms laid down in the future Protocol to the BTWC. The whole conversion process is monitored by the future BTWC Organization. The frequency and intrusiveness of investigations and monitoring decreases over the years as the conversion process proceeds.

7. References

1. See contributions by Howson, Pearson, Rimmington, Sandakhchiev and Weber in this book.
2. Meeting Report. Programme for Countering Emerging Infectious Diseases (ProCEID) by Prophylactic, Diagnostic and Therapeutic Measures. Mission Statement. Revised, in: Biologicals, Vol. 24 (1996), p. 71 - 74, see especially p. 71.
3. Text of the Convention reprinted e.g. in: Erhard Geissler and John P. Woodall (eds.): Control of Dual-Threat Agents: The Vaccines for Peace Programme. SIPRI Chemical and Biological Warfare Studies No. 15, Oxford, etc.: Oxford University Press, 1994, p. 243 - 245.
4. Special Conference of the States Parties to the Convention on the Prohibition of the Development, Production and Stockpiling of Bacteriological (Biological) and Toxin Weapons and on their Destruction. 19-30 September 1994. Final Report. Part II. Final Declaration, Geneva, 1994, p. 8 - 11, p. 10.
5. Bangladesh Statement during the General Debate at the Fourth Review Conference, 25 November - 6 December 1996, 26 November 1996, Geneva.
6. Working papers by the Friend of the Chair on Article X: "Informative Note Concerning some Activities of Multilateral Cooperation in Areas Related to the BWC and Their Relevance for Cooperation under Article X of the BWC", BWC/AD HOC GROUP/WP.23, 30 November 1995, Geneva, and "Elements for Structured Discussions on Article X on the BTWC", BWC/AD HOC GROUP/WP.98/Rev.1, September 1996, Geneva.

7. Unfortunately the Friend of the Chair group on definitions of terms and objective criteria has not yet explored, what such terms as "areas relevant to the Convention" or "areas directly relevant to the Convention" could stand for in regard to Article X measures. Such an approach would have been especially helpful for negotiations on such measures. The mandate does not limit the work of the Friend of the Chair group on definitions of terms and objective criteria to facilitating discussions in the Friend of the Chair group on compliance measures, as has so far been the case: "[T]he Ad Hoc Group shall, inter alia consider: - Definitions of terms and objective criteria, such as lists of bacteriological (biological) agents and toxins, their threshold quantities, as well as equipment and **types of activities, where relevant for specific measures designed to strengthen the Convention**; [...]" (emphasis added).
8. Meetings that do not appear in the table were allocated to other topics, technical issues or general discussions.
9. Procedural Report of the Ad Hoc Group of the States Parties to the Convention on the Prohibition of the Development, Production and Stockpiling of Bacteriological (Biological) and Toxin Weapons and on their Destruction. Annex I. Rolling Text of a Protocol to the Convention on the Prohibition of the Development, Production and Stockpiling of Bacteriological (Biological) and Toxin Weapons and on their Destruction, BWC/AD HOC GROUP/38, 6 October 1997, Geneva, p. 5 - 247, especially p. 64 - 73.
10. Activity Report 1996, International Centre for Genetic Engineering and Biotechnology, p. vii.
11. For a short description of the evolution of ProCEID see: Meeting Report. Programme for Countering Emerging Infectious Diseases (ProCEID) by Prophylactic, Diagnostic and Therapeutic Measures. Mission Statement. Revised, in: Biologicals, Vol. 24 (1996), p. 71 - 74. For more information on the earlier programme Vaccines for Peace see: Erhard Geissler and John P. Woodall (eds.): Control of Dual-Threat Agents: The Vaccines for Peace Programme, SIPRI Chemical and Biological Warfare Studies No. 15, Oxford etc.: Oxford University Press, 1994.
12. Meeting Report. Programme for Countering Emerging Infectious Diseases (ProCEID) by Prophylactic, Diagnostic and Therapeutic Measures. Mission Statement. Revised, in: Biologicals, Vol. 24 (1996), p. 71 - 74.
13. Text of the Convention reprinted e.g. in: Erhard Geissler and John P. Woodall (eds.): Control of Dual-Threat Agents: The Vaccines for Peace Programme. SIPRI Chemical and Biological Warfare Studies No. 15, Oxford, etc.: Oxford University Press, 1994, p. 243 - 245.
14. Annex. Working Paper by the Friend of the Chair. Elements for Structured Discussions on Article X on the BTWC, BWC/AD HOC GROUP/WP.98/Add.1/Rev.1, 26 September 1996, Geneva.

THE AD HOC GROUP : PAST BIOLOGICAL WEAPONS FACILITIES

GRAHAM S PEARSON
Visiting Professor of International Security
Department of Peace Studies
University of Bradford
Bradford, West Yorkshire BD 7 1 DP
United Kingdom

1. Introduction

Although Article II of the Convention on the Prohibition of the Development, Production and Stockpiling of Bacteriological (Biological) and Toxin Weapons and on their Destruction addresses the destruction or diversion to peaceful purposes of all agents, toxins, weapons, equipment and means of delivery, the conversion of past BW facilities is not addressed specifically even though such conversion is implicit in compliance with the Convention.

At the Third Review Conference [1] in 1991 the States Parties recognised the need to strengthen the Biological and Toxin Weapons Convention (BTWC) and agreed to establish an Ad Hoc Group of Governmental Experts (VEREX) to identify and examine potential verification measures from a scientific and technical viewpoint. The mandate [1] for the VEREX group required that measures be identified which could determine:

- Whether a State Party is developing, producing, stockpiling, acquiring or retaining microbial or other biological agents or toxins, of types and in quantities that have no justification for prophylactic, protective or peaceful purposes;
- Whether a State Party is developing, producing, stockpiling, acquiring or retaining weapons, equipment or means of delivery designed to use such agents or toxins for hostile purposes or in armed conflict.

The VEREX Group was chaired by Ambassador Tibor Tóth of Hungary; it met four times during 1992 and 1993 and produced a Final Report [2] which was circulated to the States Parties.

This was considered by a Special Conference held [3] on 19 - 30 September 1994 at which the States Parties *"determined to strengthen the effectiveness and improve the implementation of the Convention and recognizing that effective verification would reinforce the Convention"* agreed to establish a further Ad Hoc Group (AHG) with the objective being *"to consider appropriate measures, including possible verification measures, and draft proposals to strengthen the Convention, to be included, as*

E. Geissler et al. (eds.), Conversion of Former BTW Facilities, 201-208.
© 1998 *Kluwer Academic Publishers. Printed in the Netherlands.*

appropriate, in a legally binding instrument, to be submitted for the consideration of the States Parties." In this context, the Ad Hoc Group shall, inter alia, consider:

- *Definitions of terms and objective criteria, such as lists of bacteriological (biological) agents and toxins, their threshold quantities, as well as equipment and types of activities, where relevant for specific measures designed to strengthen the Convention;*
- *The incorporation of existing and further enhanced confidence building and transparency measures, as appropriate, into the regime;*
- *A system of measures to promote compliance with the Convention, including, as appropriate, measures identified, examined and evaluated in the VEREX Report. Such measures should apply to all relevant facilities and activities, be reliable, cost effective, non-discriminatory and as non-intrusive as possible, consistent with the effective implementation of the system and should not lead to abuse;*
- *Specific measures designed to ensure effective and full implementation of Article X, which also avoid any restrictions incompatible with the obligations undertaken under the Convention, noting that the provisions of the Convention should not be used to impose restrictions and/or limitations on the transfer for purposes consistent with the objectives and the provisions of the Convention of scientific knowledge, technology, equipment and materials.*

In addition, the Special Conference made it clear that the regime *"would include, inter alia, potential verification measures, as well as agreed procedures and mechanisms for their efficient implementation and measures for the investigation of alleged use."*

The Ad Hoc Group (AHG) under the chairmanship of Ambassador Tóth of Hungary met in January 1995 and subsequently in July 1995, November - December 1995, July 1996, September 1996, March 1997, July - August 1997 and September - October 1997. Initially, the functioning of the AHG was facilitated by the appointment of four Friends of the Chair (FOC) who chaired the sessions concerned with the four elements of the mandate:

a. Definitions of Terms and Objective Criteria - Dr Ali Mohammadi of the Islamic Republic of Iran

b. Confidence-Building and Transparency Measures - Ambassador Tibor Tóth of Hungary

c. Measures to Promote Compliance - Mr Stephen Pattison, then Ambassador Sir Michael Weston and subsequently Richard Tauwhare of the United Kingdom of Great Britain and Northern Ireland

d. Measures Related to Article X - Ambassador Jorge Berguno of Chile and then Carlos Duarte of Brazil.

At successive AHG meetings, the FOC produced papers that reflected the discussions that had taken place yet were without prejudice to the positions of delegations on the issues under consideration in the Ad Hoc Group and did not imply agreement on the scope or content of the paper. These FOC papers were considered by the AHG in plenary session and amended as requested by delegations so that they reflected the views expressed before they were accepted for attachment to the procedural reports of the meetings. It is also clear from the reports of the meetings that some FOC papers have gone through several iterations and have thus been refined and improved.

The Ad Hoc Group (AHG) has made good progress with its transition in July 1997 to negotiation of the rolling text for the legally binding instrument or Protocol. From having text for only two Articles in June 1997 the Protocol now has text for 21 of the proposed 23 Articles. [4] Although the text is liberally sprinkled with square brackets, indicating alternative views, there is actually a fair degree of consensus as to what will be included within the Protocol. It is probable that the future regime will comprise an integrated set of measures which will require mandatory declarations of facilities and activities that are of most relevance to the Convention, infrequent Non-Challenge Visits, both random and focussed, and Non-Compliance Concern Investigations together with measures to improve the implementation of other Articles of the Convention, notably Article X (international cooperation for peaceful purposes), Article III (non-transfer for prohibited purposes) and Article IV (national implementation).

There is broad consensus that the Protocol should be focussed on the activities and facilities of greatest relevance to the Convention. This paper considers how the Protocol might address one group of facilities that would generally be regarded as being of great relevance to the Convention: those facilities that were previously engaged in biological weapons programmes. Thus far, there has been no consideration by the Ad Hoc Group of the conversion of past BW facilities. The considerations in this paper are developed from the papers presented at the NATO Advanced Research Workshop (ARW) "Conversion of Former Biological Weapons Facilities: Development and Production of Prophylactic, Diagnostic and Therapeutic Measures for Countering Diseases" held at the "Frederic Joliot-Curie" Institute for Radiobiology and Radiohygiene in Budapest, Hungary on Wednesday 5 to Sunday 9 November 1997.

2. Past Biological Weapons Facilities

The programme started by addressing what might be learnt from history about the conversion of past biological weapons facilities with papers on the lessons from conversion at Porton Down in the UK and Fort Detrick in the US respectively and then considered how former facilities might be converted to work on national and international health topics. A particularly useful presentation was made by a representative of the US National Academy of Sciences (NAS) on work that they had carried out for the US Department of Defence on a possible blue-print for US - Russian collaboration to combat dangerous infectious disease as a contribution to the Cooperative Threat Reduction programme. An informal contribution was provided on the progress of the cooperative project between the US and Kazakhstan to dismantle and convert the former biological weapons facility at Stepnogorsk in Kazakhstan. Other papers in a poster session addressed topics such as the restructuring of the State Research Center of Virology and Biotechnology Vector and diagnostic kits that could be produced for bacterial and viral diseases.

The aim of the NAS study was to design a program to expand bilateral cooperative efforts between US scientists and their Russian counterparts who had been involved in the former Soviet BW program. National security benefits identified included providing greater confidence and transparency about activities at former Soviet BW facilities, reducing proliferation incentives and providing a stepping stone to dismantling. It was, however, noted that even the most extensive collaboration between Russian and US scientists would not provide incontrovertible assurance that all research activities on dangerous pathogens are devoted to legitimate purposes. Recognising this risk the NAS committee had concluded that if carried out under appropriate rules of transparency, a cooperative program could be carried out in a manner that ensures the risk of abuse of such a programme is reduced to an acceptable level. Arrangements to provide sufficient transparency should include mutual agreement on a project by project basis of activities that are legitimate, regular and agreed-upon access to facilities, personnel and information; and commitment to the principle that providing assurance is an active rather than a passive responsibility. Moreover, during the evaluation of any proposed joint research project, its potential contributions to health and national security must be judged to outweigh the risk that the project might contribute to development or improvement of offensive BW activities.

The contribution on the Stepnogorsk dismantlement project emphasised that Kazakhstan had been completely open about both past and present activities at Stepnogorsk; no comparable openness had yet been achieved with the Russian Federation. The Stepnogorsk facility had been set up to produce anthrax and to fill this into munitions. Although test runs had been carried out the plant had been intended to be used for full production during mobilization. Several million US $ were being spent in a dismantlement and conversion project. It was noted that in this conversion most of the containment was being removed. It was said that the cooperation with Kazakhstan was going very well, that there was tremendous transparency and that a real sense of trust was being achieved.

The general sense that appeared from these presentations concerning both former Soviet Union as well as UK and US past BW facilities was that such facilities could present a risk to the Convention as their staff had the essential knowledge about how to develop and produce BW. There was thus a real need for complete transparency and openness in such past BW facilities. Consequently the aim should be within such facilities to dismantle those capabilities that have no potential peaceful applications (such as hardened filling and storage buildings) and to convert the facilities to work on countering public health concerns. It was, however, recognised that the first step could be to convert such facilities to biological defence activities; such converted facilities could, however, relatively easily revert to prohibited BW activities. Steps needed to be taken to build confidence that there were no such reverted activities.

3. CWC Provisions for Chemical Weapons Production Facilities

The ARW also took particular note of the current state of play of the negotiations of the Ad Hoc Group of the States Parties of the Biological and Toxin Weapons Convention (BTWC) which is considering measures to strengthen the Convention through a legally binding instrument. The ARW was greatly aided in this by a presentation on the provisions in the Chemical Weapons Convention (CWC) for Chemical Weapons Production Facilities (CWPFs). The CWC makes it clear that CWPFs, in the narrow sense in which the treaty defines such facilities, should be destroyed although they can, with the agreement of the OPCW, be converted into Chemical Weapons Destruction Facilities and then destroyed. In addition, a State Party may request, in exceptional circumstances, permission from the Conference of States Parties for a CWPF to be used for purposes not prohibited under the CWC. The CWC in Article V states that:

8. Each State Party shall destroy all chemical weapons production facilities .... Such destruction shall begin not later than one year after this Convention enters into force for it, and shall finish not later than 10 years after entry into force of this Convention. ...

13. A State Party may request, in exceptional cases of compelling need, permission to use a chemical weapons production facility ... for purposes not prohibited under this Convention. Upon the recommendation of the Executive Council, the Conference of the States Parties shall decide whether or not to approve the request and shall establish the conditions upon which approval is contingent ...

14. The chemical weapons production facility shall be converted in such a manner that the converted facility is not more capable of being reconverted into a chemical weapons production facility than any other facility used for industrial, agricultural, research, medical, pharmaceutical or other peaceful purposes not involving chemicals listed in Schedule 1.

15. All converted facilities shall be subject to systematic verification through on-site inspection and monitoring with on-site instruments...

Detailed conditions for conversion to purposes not prohibited are specified in Part D of Annex V of the Verification Annex of the CWC. These include the requirement that should this permission be granted then the converted facility shall not be used for any activity involving a Schedule 1 or Schedule 2 chemical or for the production of any highly toxic chemical and, in addition, for 10 years after conversion, the facility shall be subject to a particularly intrusive verification regime:

"For the 10 years after the Director-General certifies that conversion is complete, the State Party shall provide to inspectors unimpeded access to the facility at any time. The inspectors shall have the right to observe all areas, all activities, and all items of equipment at the facility. The inspectors shall have the right to verify that the activities at the facility are consistent with any conditions established under this Section, by the Executive Council and the Conference. The inspectors shall also have the right...to receive samples from any area of the facility and to analyse them to verify the absence of Schedule 1 chemicals, their stable by-products and

decomposition products and of Schedule 2 chemicals and to verify that the activities at the facility are consistent with any other conditions on chemical activities established under this Section, by the Executive Council and the Conference."

After 10 years, the Executive Council will decide on the nature of continued verification measures. These CWC requirements in respect of CWPFs clearly reflect the risk to the Convention posed by such facilities if they are not destroyed. Furthermore, they reflect the fact that Schedule 1 have little if any civil use and Schedule 2 chemicals are used commercially in relatively small quantities and hence the likelihood of a requirement for such facilities for peaceful purposes is slight.

4. Ad Hoc Group

In considering past BW facilities, the ARW noted that Article II of the BTWC requires each State Party to undertake *"to destroy, or to divert to peaceful purposes, as soon as possible....all agents, toxins, weapons, equipment and means of delivery specified in Article I of the Convention..."*. There is no specific mention in this Article or elsewhere in the BTWC of biological weapons facilities and what should happen to these under the Convention although it is clearly intended that there should be no continued requirement for such facilities. As the AHG is considering measures to strengthen the BTWC, it would clearly be appropriate and relevant to consider measures in respect of biological weapons facilities.

The following ideas about measures relating to past biological weapons facilities were developed at the ARW which recognised that because of the dual purpose nature of biological and toxin agents, there was likely to be a requirement to retain the dual purpose capabilities for peaceful purposes although weapons related equipment should be destroyed. The enhanced CWC verification requirements for converted CWPFs reflected the risk to the CWC from such facilities. Because the nature of BW is different from that of CW and because of the practicality of producing significant quantities of BW agents in R & D facilities, it follows that the risk to the BTWC is from BW facilities and is thus broader than that from BW production facilities alone. It was also noted that in the Working Papers presented at the AHG, some States had proposed that some facilities, such as those engaged in military biological defence programmes, might be subject to more frequent inspections. Consequently, there would be a logic in the AHG adopting an intensified verification regime for converted BW facilities; the degree of this intensification might usefully reflect the extent to which the facility was still engaged on work with pathogens and toxins generally regarded as BW agents and was under government control. After a period of time, such as 10 years, assuming that the facility was no longer engaged in work with BW agents and was no longer under government control, the verification regime might be relaxed to one that is the same as that for other declared facilities.

It was noted that the current rolling text developed by the AHG includes provision for the declaration of past offensive BW programmes (Article III. D and Appendix C).

Although Article III.D usefully seeks, albeit in square brackets, a comprehensive declaration on "past offensive...biological research [and] development [testing and production] programmes", Appendix C in the current draft simply reiterates the language agreed for Confidence-Building Measure (CBM) F at the Third Review Conference in 1991. This requires States parties to provide a "Summary of the research and development activities indicating whether work was performed concerning production, test and evaluation, weaponization, stockpiling of biological agents, the destruction programme of such agents and weapons, and other related research." The CBM requires declaration of past offensive programmes conducted since 1 January 1946.

Elsewhere in the rolling text, although there is currently no text for "Programmes and Facilities" in Annexe A.V on Declarations, language is provided in Appendices A and B for information to be included in declarations of biological defence programmes and facilities taking part in biological defence programmes. It would be reasonable and logical to include in Article III.D the requirement for declarations of converted BW facilities and for the modalities in Annexe A.V and Appendices A and B to require information on the programmes and facilities of converted BW facilities. It would be consistent with the already agreed CBM for the rolling text to require such declarations for all BW facilities converted since 1 January 1946.

In addition, the rolling text has in Article III.F language for visits and investigations. The categories of visits currently include Random Visits, Ambiguity-Related Visits, Clarification Visits, Request Visits and Voluntary Visits and it would appear reasonable to include a further category of Visits to Converted Facilities which might occur more frequently under a broader mandate than the other proposed Visits.

5. Conclusion

The subject of measures to be adopted in a Protocol to strengthen the BTWC in respect of converted BW facilities has not been directly addressed by the AHG thus far. It is clear from consideration of the provisions of the CWC that past CWPFs have been explicitly addressed in that Convention and, should, exceptionally, such facilities be converted, they will then be subject to a more intrusive verification regime. The considerations at the NATO ARW led to the view that converted past BW facilities likewise need to be explicitly addressed in the legally binding instrument being negotiated by the AHG in Geneva. Appropriate measures could readily be incorporated in the language on declarations and on visits being assembled in the current rolling text. There is a clear opportunity now for the AHG to address this in their negotiations and one that should not be missed.

6. References

1. United Nations, *The Third Review Conference of the States Parties to the Convention on the Prohibition of the Development, Production and Stockpiling of Bacteriological (Biological) and Toxin Weapons and on their Destruction*, Geneva, 9–27 September 1991, BWC/CONF.III/23, Geneva 1992.

2. United Nations, *Ad Hoc Group of Governmental Experts to Identify and Examine Potential Verification Measures from a Scientific and Technical Standpoint*, Report BWC/CONF.III/VEREX/9, Geneva 1993.
3. United Nations, *Special Conference of the States Parties to the Convention on the Prohibition of the Development, Production and Stockpiling of Bacteriological (Biological) and Toxin Weapons and on their Destruction*, Final Report, BWC/SPCONF/1 Geneva, 19–30 September 1994.
4. United Nations, *Ad Hoc Group of the States Parties to the Convention on the Prohibition of the Development, Production and Stockpiling of Bacteriological (Biological) and Toxin Weapons and on their Destruction*, BWC/AD HOC GROUP/39, Geneva, 2 February 1998.

PROPOSED GLOSSARY OF TERMS AND DEFINITIONS

This glossary is the result of extensive discussions on individual terms that are used in the contributions of this book. It is not intended to provide universally valid definitions but rather to reflect preliminary work on these terms in the context of BTW facilities' conversion. In the process of working on the specific topic of BTW facilities' conversion these terms have to be further developed and adapted according to the existing literature on the broader conversion subject.

BTW facility[1]	A facility engaged in activities prohibited under the BTWC, i.e. developing, producing, stockpiling, or otherwise acquiring or retaining microbial or other biological agents or toxins whatever their origin or method of production, of types and in quantities that have no justification for prophylactic, protective or other peaceful purposes, or weapons, equipment or means of delivery designed to use such agents or toxins for hostile purposes or in armed conflict.
BTW defence facility	"Facility which works in [one or more of the following areas of] [a biological defence programme] [/ defence programme against biological and toxin weapons] [as one of its principal and/or permanent roles in research, development, testing, production and evaluation]."[2]
Closure of a BTW facility	The process of stopping all activities at a BTW facility by removal of staff, but without (immediate) destruction, dismantling or conversion.
Conversion of a BTW facility	The process of transforming a BTW facility into a facility pursuing activities that are not prohibited under the BTWC. Distinction should be made between conversion into a BTW defence facility (partial conversion), or a facility that is engaged exclusively in civilian activities (full or total conversion).

Destruction of a BTW facility	The process of razing a BTW facility to the ground, ensuring that neither the whole facility nor components or equipment can be used again. Demolition is deliberate, calculated destruction.
Dismantling of a BTW facility	The process of taking apart a BTW facility in order to make impossible the functioning of this facility as a BTW facility. Equipment, however, can be used in other places for other purposes. Distinction should be made between taking away all the equipment (full dismantling) and taking away only equipment characteristic for offensive BTW activities (partial dismantling).
Reversion	The process of transforming a converted former BTW facility into an active BTW facility.
Cessation	The process of bringing activities at a BTW facility to an end by closure, conversion, dismantling or destruction.

Endnotes

1 The term "facility" is understood according to the definition provided in the rolling text of a Protocol to the BTWC (BWC/AD HOC GROUP/38, 6 October 1997, p. 18): "A combination of physical structures, equipment, workforce and principal associated support infrastructure [having an identifiable boundary and a single administration] whether under construction, operational or non-operational. [for [the] [either] [research,] development, production, testing, processing, stockpiling, otherwise acquiring or retaining microbial or other biological agents or toxins]." Accordingly, a facility can be as small as a few rooms in a bigger building or as big as a large building complex.

2 The definition provided here copies the definition of the term "Biological defence facility" in the rolling text of a Protocol to the BTWC. BWC/AD HOC GROUP/38, 6 October 1997, p. 17.

ACRONYMS

AAI	Allen & Associates International Ltd.
ACDA	Arms Control and Disarmament Agency
AHG	Ad Hoc Group
AO Progress	Joint Stock Company Progress
ARW	Advanced Research Workshop
BDP	Biology Department Porton
BSE	Bovine spongiform encephalopathy
BW	biological weapons
BWC	Biological Weapons Convention
CAMR	Centre for Applied Microbiology and Resaerch
CBDE	Chemical and Biological Defence Establishment
CBMs	Confidence Building Measures
CBW	Chemical and Biological Warfare
CD	Conference of Disarmament
CDC	Center for Disease Control and Prevention
CDE	Chemical Defense Establishment
CFR	Code of Federal Regulations
CGIAR	Consultative Group on International Agricultural Research
cGMP	current Good Manufacturing Practice
CIS	Community of Independent States
CISAC	Committee on International Security and Arms Control
CRDF	Civilian Research and Development Foundation
CTBC	Comprehensive Test Ban Treaty
CTR	Cooperative Threat Reduction Program
CW	Chemical Weapons
CWC	Chemical Weapons Convention
CWDF	Chemical Weapons Destruction Facilities
CWPFs	Chemical Weapons Production Facilities
DOD	Department of Defense
DPT	Diphteria-pertussis-tetanus
DSWA	US Defence Special Weapons Agency
DTAs	Dual-Threar Agents
ECCO	European Culture Collection Organization
EIF	entered into force
EPI	Expanded Programme for Immunization
EU	European Union
FAF	Federal Armed Forces
FCRDC	Frederick Cancer Research and Development Center
FDA	Food and Drug Administration
FFCD	Full, Final and Complete Disclosure
FMCT	Fissile Material Cut off Treaty
FMD	Foot-and-mouth disease

FOC	Friends of the Chair
GMP	Good Manufacturing Practice
HBV	Hepatitis B Virus
HDF	Hungarian Defense Forces
HEPA	High Efficiency Particulate Air
HuMAbs	Human monoclonal antibodies
HVAC	Heating, Ventilation and Air Conditioning
IAGGI	Independent Advisory Group on Gruinard Island
ICGEB	International Center for Genetic Engineering and Biotechnology
IPP	Initiatives for Proliferation Prevention
IPR	Intellectual Property Rights
IQ	Installation Qualification
IRRI	International Rice Research Institute
ISF	International Science Foundation
ISTC	International Science and Technology Center
KWI	Kaiser-Wilhelm-Institut
MMA	Military Medical Academy
MOD	Ministry of Defense (Russian Ministry of Defense)
MRD	Microbiological Research Department
MRE	Microbiological Research Establishment
NAS	National Academy of Sciences
NASA	National Aeronautics and Space Administration
NBC	Nuclar Biological Chemical
NCAs	National Control Authorities
NCI	National Cancer Institute
NGO	Non Governmental Organizations
NIH	National Institutes of Health
OMV	Ongoing monitoring and verification
OPV	Oral polio vaccine
OQ	Operational Qualification
OTS	Operations and Technical Support
PQ	Performance Qualification
ProCEID	Programme for Countering Emerging Infectious Diseases
ProMED	Programme for Monitoring Emerging Diseases
QA/QC	Laboratory Quality Assurance/Control Laboratory
QTLs	quantitative trait loci
R&D	Research and Development
SAIC	Science Applications International Corporation
SOPs	Standard operating procedures
SRC	Scientific Research Center
SRCVB	Scientific Research Center for Virology and Biotechnology
STCU	Science and Technology Center Ukraine
TACIS	Technical Assistance to the Community of Independent States
TOP	Turnover Package

TW	Toxin Weapons, Toxin Warfare
UK	United Kingdom
UN	United Nations
UNSCOM	United Nations Special Commission
USAMRIID	United States Army Medical Research Institute for Infectious Diseases
USDA	U.S. Department of Agriculture
VEE	Venezuelan equine encephalomyelitis
VEREX	Ad Hoc Group of Governmental Experts
WFI	Water For Injection
WMDIE	Weapons of Mass Destruction Infrastructure Elimination
WWI	World War I
WWII	World War II

INDEX

MIX
Papier aus verantwortungsvollen Quellen
Paper from responsible sources
FSC® C105338
FSC
www.fsc.org

If you have any concerns about our products,
you can contact us on
ProductSafety@springernature.com

In case Publisher is established outside the EU,
the EU authorized representative is:
Springer Nature Customer Service Center GmbH
Europaplatz 3, 69115 Heidelberg, Germany

Printed by Libri Plureos GmbH
in Hamburg, Germany